On Time
Technology
Implementation

**How to Achieve
Implementation Success
with Limited Time and
Resources**

On Time Technology Implementation

How to Achieve Implementation Success with Limited Time and Resources

Bennet P. Lientz
University of California, Los Angeles
Los Angeles, California

Kathryn P. Rea
The Consulting Edge, Inc.
Beverly Hills, California

Routledge
Taylor & Francis Group

LONDON AND NEW YORK

First published by Academic Press

This edition published 2011 by Routledge
2 Park Square, Milton Park, Abingdon, Oxon OX14 4RN
711 Third Avenue, New York, NY 10017, USA

Routledge is an imprint of the Taylor & Francis Group, an Informa business

Library of Congress Catalog Card Number: 99-63089

International Standard Book Number: 0-12-449975-9

Contents

Part I
Set the Direction

Chapter 1
Introduction

v

Chapter 2
Develop the Business Process Plan and Strategy

Chapter 3
Defining the New Process, Benefits, and Requirements

Chapter 4

Assess Outsourcing

Part II
Software Development

Chapter 5
Define Requirements and Do Prototyping

Chapter 6
Perform System Development and Integration

Chapter 7
Conversion, Procedures, and Training

Chapter 8
Perform Maintenance, Enhancement, and Production Support

Part III
Software Packages

Chapter 9
Carry Out Evaluation and Selection

Chapter 10

Conduct Negotiations and Define the Implementation Plan

Chapter 11

Implement the Package

Part IV
Knowledge Management

Chapter 12

Identify Opportunities and Develop Your Strategy

Chapter 13

Analyze Knowledge Management Technologies and Services

Chapter 14
Implement Knowledge Management

Part V
Electronic Commerce and Implementation Issues

Preface

INFORMATION TECHNOLOGY FAILURES AND PROBLEMS

The 50th anniversary of trying to cope with computer technology and software is approaching. Hardware and communications have made tremendous strides. Software tools and aids have also improved but at a slower pace. People are supposed to be smarter because of the half-century of experience, right? Well, you would not think so if you pick up almost any information technology magazine. There are frequent stories of disasters. Here are some examples:

- Before the mid-1940s there were more than $250 billion in software projects in the United States spread across about 175,000 projects. It was estimated that there were more than $59 billion in cost overruns and another $81 billion in canceled projects.
- More than one half of the client/server projects either fail or experience major cost overruns.
- More than one third of the attempts to implement major software packages fail.
- Almost one half of reengineering and process improvement projects fail.
- A company attempted to implement a major integrated software system and failed after spending more than $10 million. There was nothing to be salvaged. People on the project thought they were doing fine. Obviously, they were not.
- Electronic commerce is one of fastest growing areas of information technology (IT). Yet, companies are making the same mistakes here as with other systems.

In rating reasons for problems, a survey revealed the following as the top reasons in order:

1. Lack of top management commitment
2. Failure to gain user commitment
3. Misunderstanding requirements
4. Lack of user involvement
5. Failure to manage end-user expectations
6. Changing scope and objectives
7. Lack of required knowledge and skills in the team
8. Changing requirements
9. New technology
10. Insufficient or inappropriate staffing
11. Conflicts among user departments

Why do such problems and disasters persist? Why do people not learn? There are some common myths that seem to contribute to recurring failure. One is that each situation is totally unique. People sometimes claim that they are too busy to learn or that technology is changing so much that experience has little value. Neither of these statements is true. Failures recur because traditional failed technical and management methods continue to be used.

PURPOSE AND SCOPE OF THE BOOK

This book addresses each of the previously mentioned problems in detail. A basic premise of the book is that the business and technology environments around a system are never static. There are really no such things as frozen requirements. Systems are closely linked to the business processes they support. As such, both are beset by changing management, new government regulations, impacts from other processes and systems, new technologies, and improved methods and tools.

This book helps you to implement systems and technology successfully, with limited resources and time. Step-by-step proven methods are provided for development, software packages, and knowledge management.

Many critical systems activities in the recent past and foreseeable future involve the following areas:

- Implementing improved business processes through technology
- Designing and developing software
- Implementing software packages
- Managing changes to existing software through maintenance and enhancement
- Implementing knowledge management solutions
- Establishing and managing consulting and outsourcing work

These six areas constitute the scope of the book. These areas are where a large percentage of systems and technology implementations fall apart. They are also the ones that consume the resources and deliver the benefits to the business.

WHAT YOU WILL GET OUT OF THE BOOK

There are many books on each of the major topics identified here. Why write another implementation book? Some of the reasons are as follows:

- Many of the existing books are written from an academic view with little relation to the real world. In fact, many of the techniques employed in such books either are no longer used in industry or have little value.
- Other books are written with a bias toward a specific method or tool. This is nice if you are going to use the tool but of limited value otherwise.
- There is a shortage of lessons learned from actual implementation experience—success, failure, and everything in between.
- Finally, many books are based around the classical old sequential life cycle of systems, developed in the 1950s for the technology of that period and a direct contributor to failures of today.
- Most books are written as if implementing systems were detached from real world business processes. Yet, the processes are where the benefits and justification of the systems and technology are.

This is a "how to" book, not a "what" book. Who can use the book? Anyone in business and IT who is involved in developing systems, implementing software packages, and modernizing critical business processes can use this book. The materials and guidelines here have been developed, tested, and used since the mid-1980s in more than 50 organizations of all sizes, shapes, industries, and locations. They have been employed in organizations of different sizes, locations, and industries. There is no requirement to have a technology background or in-depth technical knowledge.

What should you get out of this book? Here are some critical features:

- How to implement electronic commerce solutions
- How to implement Enterprise Resource Planning (ERP) software
- How to cope with and manage maintenance, enhancement, and production support
- How to evaluate, select, and implement software packages
- How to implement knowledge management, lessons learned, and data warehousing
- How to develop modular, maintainable software
- How to control changing requirements and scope

- How to involve business departments to increase the odds of success
- How to apply techniques in intranet, concurrent engineering, and collaborative project management as well as finish work on time and with reduced cost

There are more than 200 guidelines to address many of the problems and situations that you will face. In addition, there are checklists and examples supporting the methods presented. The style of the book is informal and common sensical. Technology alone can be dry, but when you add a dry and boring writing style to it, the result can be disaster.

What should you hope to accomplish by employing the techniques detailed here? Tangible results have been solving problems, reducing elapsed work time, controlling costs, and improving the quality and benefits of the result. You should be able to build a more collaborative and cooperative approach among systems staff, business staff, and vendor staff.

THE APPROACH OF THE BOOK

From experience, critical success factors for technology implementation are based on the following themes:

- Successful technology implementation occurs only if the business process is improved. Costs lie in the technology and support. Bottom-line business benefits lie in the process.
- Business departments must play the leading role in implementation. This stabilizes requirements, gains business commitment, and better ensures that benefits in the process will be achieved.
- To achieve results on time and within budget, you must implement and perform work concurrently. A sequential approach is doomed.
- Modern, collaborative project management that addresses issues, risk, and change control is critical to implementation success.

Now look at the title of the book. The key phrase there is "on time." Why aren't "cost effective" or "successful" used? Experience shows that throwing money at implementing technology can often make the situation worse, not better. You can always find more money for worthwhile projects, but you cannot buy time. Experience with more than 70 projects have revealed that the consequences of extending time in implementation can result in disaster. Confidence can be lost. The direction of the project can be forgotten. The goals become lost.

As you may be aware, this book is one of several other books on project management and process improvement, including *Project Management for the 21st Century* (2nd edition, Academic Press) and *Breakthrough Technology Project*

Management (Academic Press). This book complements these two books by focusing on the how to do the technical part of systems and technology projects.

ORGANIZATION OF THE BOOK

The book is divided into five parts:

- Part I sets the direction for implementation. Here you select the critical business process and system to be addressed, define a new business process, and set the course toward development, software packages, or knowledge management.
- Part II addresses software design, development, conversion, and maintenance and support.
- Part III takes you step by step through evaluating, selecting, planning, and implementing software packages.
- Part IV is concerned with implementing knowledge management applications, including lessons learned, data warehouses and marts, and data mining.
- Part V is a summary that addresses electronic commerce and how to deal with some of the major issues in implementation.

Each chapter contains an introduction, core chapter material, key areas where things can go wrong, tips on reducing the schedule and costs, suggestions for lessons learned, a summary, and steps you can take after reading the chapter. There are a number of examples, including central examples from manufacturing, banking, insurance, transportation, and government.

Bennet P. Lientz
Kathryn P. Rea

Part I

Set the Direction

Chapter 1

Introduction

You are involved in implementing technology or a system, You want to be successful. To be successful, you need to complete the implementation on time and within budget. You also want the systems and technology to be used successfully in a business process. This is the definition of success in this book—implementing a successful system and process, and doing so within the constraints of time and money.

INTRODUCTION

Experience has shown that time is often more important than money. Although a manager can throw money at a project, there is only a certain amount of elapsed time. Moreover, there are many stories to indicate that putting more money into a project can only make matters worse. Applying money means adding resources. Adding hardware or software is more easily handled; however, adding more people can be a disaster because they slow the progress of the current staff as they learn what is going on. They may also question what has already been done. Time marches on and the work gets further behind. In one instance, a manager wanted to end an implementation project. It was politically impossible to do it, but it had to be done because the business had changed and the need for the system evaporated. What to do? It was recommended that the manager add people. The project died in less than 1 month. This is an extreme example, but it reinforces the point. Negotiating for more time makes sense; trying to get more money often makes less sense.

To provide implementation guidelines, trends in both business and technology need to be considered. Next, some of the methods of the past can be viewed in light of current and future trends. This leads to the proven approach that has worked in more than 50 organizations around the world.

3

BUSINESS DIRECTION OF INFORMATION TECHNOLOGY

It is a cliché to say that business depends on technology more today than ever before. Thus, we examine some aspects of the interdependence between the business and systems and technology. In the early years of computer technology, business received benefits by reducing the extent of clerical and support personnel required to perform routine tasks. Businesspeople often viewed technology at this stage as a necessary, but largely unknown and mysterious, expense. With the arrival of stand-alone PCs, the mystery disappeared, along with a lot of money. The problem was that it was very difficult to justify the hardware and support costs of stand-alone PCs for many employees. Data were reentered from mainframe and minicomputer reports into PCs. People would show up at meetings with contradictory data that they had entered. In one case, it was recommended to a distribution firm that almost all stand-alone computers be confiscated and warehoused. When this occurred, productivity immediately improved and costs were reduced—a quick payoff achieved by reducing technology.

The trend was reversed with the development of local and wide area networking and, especially, with the Internet. Employees could perform online transactions through client-server systems. Data were made available for analysis through database management systems, and information could be shared among staff. The Internet and the World Wide Web produced even more dramatic change. First, widespread remote operations could be inexpensively linked with headquarters units. Second, customers and suppliers could be integrated. The use of electronic commerce and its predecessor, Electronic Data Interchange (EDI), has increased dramatically.

What does this add up to? It translates into the realization of the dependence of business on technology. It also raises the performance bar in terms of what must be accomplished by systems and technology. The standards have been raised. The system or technology not only must work, but also must mesh with the business and produce concrete results. Firms are also more aware of disaster stories and the problems with poor or faulty implementation. Two large discount chains implemented the same technology over a period of years—satellite communications with stores, point of sale (POS), scanning of bar coding, EDI, shipping container marking, dynamic inventory, and so on. In both cases, the systems worked. However, in one there was tremendous business success. In the other, it was a disaster. Why the difference? Because the successful firm implemented tight integration between the business processes and technology. The processes were changed to take advantage of the technology. In the other firm, the processes remained the same. The first firm made more money, had the flexibility to expand faster and cheaper, managed inventory more tightly, and could target specific market segments. The second firm received little benefit. The lessons learned here are that technology implementation is too important to be treated as a technical project and, to gain benefits processes and systems must be integrated.

BUSINESS PROCESSES AND TECHNOLOGY

Businesses depend on their critical business processes. A *business process* is a set of interrelated activities that are performed to process specific work in the business. Payroll, marketing, warehousing, and accounting are all business processes. Organizations make or lose money depending on their processes. This was recognized in ancient Rome where military and civil work was systematically reduced to formal procedures. Business processes were created and expanded throughout the Renaissance and during the Industrial Age, with the railroads and modern logistics. Industrial engineering and "efficiency experts" addressed steps in business processes in excruciating detail (e.g., hand movements in performing work). From the 1950s to the 1970s, business was preoccupied with expansion and other activities, and viewed computers and technology as support to business processes. The clumsy technology could in no way restructure the business process; it could only replace some steps.

This changed with online systems in the 1980s. Transaction-based systems changed the way people did their work. From administrative online systems to automated teller machines (ATMs) to voice response systems, business processes changed with the technology. The trend has continued with client-server and intranet/Internet systems. Currently, most critical business processes in a company are tightly connected to systems. In addition, it is hard to think of systems that do not support business processes. Yet, where do things go wrong? It is a matter of degree. If you spend money, time, and resources on critical business processes, then you will be more successful. However, if resources are drained to support minor maintenance and enhancement changes that have little or no strategic impact, both money and time are wasted. The promise of technology goes unmet.

Another major theme is that business processes such as technology evolve. As time goes by, new situations arise. If the current process cannot handle them, then people may create new, little business processes on the side. These are called *shadow systems*. They may be procedures or PC systems that are not part of the formal process. Shadow systems act to undermine the integrity of the process. The situation worsens when there is turnover of business staff and the new staff is not properly trained in the new process. New employees are told to take their past experience and knowledge and just plunge in. And you wonder why processes can deteriorate? A good combination of technology and process reinforce each other and stave off, or at least delay, decay and rot.

Failure can occur if a firm adopts new technology too soon. An example is a retail firm that adopted their own coding system for their merchandise, prior to the development of bar coding. When bar coding became popular, the firm had to use both the internal code and the bar code. The systems complexity overwhelmed the business process, and inventory controls collapsed. Some stores had excessive merchandise; others had severe shortages of the same goods. However, if you adopt technology too late, you can lose in competition. Timing is essential.

TECHNOLOGY AND SOFTWARE TOOL TRENDS

People think of technology and its use as constantly improving and expanding. However, this is not necessarily true. Much of new technology has to be hyped as "products in search of problems to solve." The electric light bulb was that way in the 1800s. It took years to for Thomas Edison to convince people that electric lights were a good idea. Do you realize today that almost one half of the world's population has never used an electric light? Getting technology adopted is often more of a marketing problem than a technical problem.

Technology and, in particular, software have experienced positive trends that relate to business and business processes. These include the following:

- *Improved price performance and reliability.* You have heard about this for years. However the changes are now more integrated, occur faster, and have a greater impact on society and individuals. Look at the sharp rise in tools that use the Internet for voice communications and faxing.
- *Greater standardization.* With the earlier dominance of several hardware firms and now with several software firms (one in particular), there is much more standardization than was previously the case. What does standardization mean? It means that you can more easily implement systems and have less maintenance later.
- *Integration and easier interfaces.* This relates to standards in that the rise of ODBC, OLE, and ActiveX have all supported simpler interfaces and integration among systems. Without this there would be a need for much more custom programming and development to establish and support interfaces.
- *Reduced learning curve for professionals to come up to speed.* Yes, current tools require learning time. However, if you compare today's tools with the tools of the past (assembly language, APL, COBOL, etc.), you find that the

new tools are more modular and fully functional. People develop basic skills and generate useful software more quickly, making them more productive and providing a more nourishing environment.

CHALLENGES FACING INFORMATION TECHNOLOGY AND MANAGEMENT

Information technology (IT) faces a number of implementation challenges — in part, generated by its success. Some challenges are addressed as follows:

- **How to deal with rising expectations of management for benefits from technology.**
  Similar to the classic "child in a candy store," once people realize benefits from technology they want more. Listening to vendor marketing pitches and advertising, it seems as simple as picking up the telephone. Obviously, it is not. So what should you do? You will be including the change of the business process within their expectations. You will find that this will shift considerable management attention away from the technology and toward the business process.
- **How to successfully involve business unit staff in implementation work.**
  Business people have to be motivated to become involved and stay involved on their own. If you force them into implementation, it will not be effective. Therefore, you will be provided with guidelines for getting and keeping people involved in useful work.
- **How to implement new technology with changed processes on a timely basis.**
  You can benefit from guidelines not only for what to do, but also for what not to do. What are some of the steps that you could delete? As you will see, this is a series of trade-offs. For example, how much and what documentation should be produced is a trade-off between benefit and effort. It is similar for other activities.

CRITICAL SUCCESS FACTORS IN SYSTEMS AND TECHNOLOGY

Following is a list of some critical factors that you must consider in implementation:

- The benefits of systems and technology implementation lie for the most part in the business process. The only benefits that lie in IT are cost savings through technology replacement and efficiency. Because IT costs are small in comparison with business costs and IT support costs for a business process are small in comparison with that of the process, it is natural to go to the process rather than IT for benefits.

- Given the cost and time of implementation, select the processes and systems based on overwhelming benefits. If the tangible benefits are very close to costs, then any rise in costs and extension of schedule will eat up the benefits. The exception is when you are implementing a major upgrade in technology infrastructure that will pave the way for major benefits in several business units.
- The scope of implementation must include commitment by the business units as well as the business process. If this is not done, then there will be few guarantees that the benefits will occur because there will be little pressure on business units to make changes. In some cases, middle-level managers want to protect their territories and realms.
- Technology that is proven should be the technical base of the implementation. In addition, the importance of interfaces and integration can outweigh the internal features of the software.
- Selection of software packages should be based on interfaces, fit with transactions of the new business process, modularity as opposed to checklists of features, and ability to customize the package. Ask the question, "Which package fits the new process best?" Do not ask "Which package has the most features?"

TRENDS IN SOFTWARE DEVELOPMENT

Traditionally, requirements were developed. After these were signed off, the system was designed. After the design was approved, the system was built, tested, and integrated—the traditional life cycle. How can this process be different with modern technology? With the widespread use of Windows-type software and Web browsers, user interfaces are more standard. Almost every new system you implement must interface and integrate with other, existing systems, further restricting what can be done. Some things, such as user reports, are now obtained through files sent to spreadsheets. Development tools are more modular, and there are software libraries.

So where is the creativity? What is the new process of development? Creativity lies in how you design the new business process through transactions with the technology and process. Instead of performing the entire analysis, design, and development, you pursue a parallel approach, where groups of transactions are in various stages of development at any one time. This allows for reuse of the design and code from earlier transactions.

TRENDS IN SOFTWARE PACKAGES

Software packages have lagged behind the software tools. This is to be expected because software vendors must rewrite their systems using the new tools. This occurred with the vendors' adoption of database management systems, fourth-generation languages, client-server systems, and intranet systems. Packages now have flexibility through control tables and, to some extent, through object libraries. The next wave of change in packages may be toward object-oriented software packages in which an organization customizes an application much more than is possible with tables.

The popularity of software packages is growing for several reasons. The packages have more capabilities than in the past. Companies lack the internal resources and time to develop the software on their own. The internal legacy systems have aged to the point where many must be replaced (e.g., Year 2000 problems). Management views these packages as a rapid fix to both business processes and systems—just shove the package and fit the process around the package. Failure occurs often because of this last faulty assumption. Processes are not rubber bands or silly putty that can fit anything. Instead, they reflect the unique organizational, cultural, social, political, and industrial settings of the firm at that time and place.

In the past, you could select a package and then modify it to fit your current process. Many vendors now resist this for several reasons. First, customization brings problems in keeping new versions of the software compatible with various customized efforts. Second, vendors only make profits in overhead on time and materials for customization. It is often more fruitful to devote these resources to new releases and versions of the core software.

TRENDS IN KNOWLEDGE MANAGEMENT

Knowledge management was once viewed as consisting of data warehousing and data marts, exclusively. The idea was that you gathered data into a warehouse and then used fairly exotic software tools perform data mining. The concept was that this would help to give you a competitive edge. For some firms, this has worked successfully; for others, it has been a disaster. No one used the data in some firms and the data were not valid or useful in others. Today, knowledge management is wider in scope. It encompasses gathering lessons learned and making this information available to others. It also means gathering information on how to do work in a business process more effectively and getting employees to use the experience across time and space.

Knowledge management represents a unique area separate from software development and software packages that target operational applications.

Knowledge management software tends to have an impact on the more strategic processes of an organization as opposed to standard software.

TECHNIQUES OF THE PAST

The early days of programming were complex and involved massive amounts of labor that yielded little results when compared with those of today. Many readers of this book are probably unfamiliar with what it was like. Although you can imagine punched cards, batch processing, and time delays for getting results from testing, it is difficult to imagine the pitiful editing, debugging, and testing tools. Hours were spent reading a hexadecimal memory dump to see what happened with a specific program that blew up. Although COBOL and FORTRAN helped, the problems persisted. This environment clouded the approach that was taken in implementing systems.

First, there had to be a life cycle. That is, you had to nail down requirements. That was because if you allowed later changes in requirements, the design and programming would unravel. Months of work could be lost. So business units had to sign off in blood that requirements were final—even if they did not understand them. Similarly, for purposes of control, the design had to be completed, and signed off by the users prior to programming. Aspects of the design sometimes could not be implemented due to the limitations on the technology.

A second impact was on documentation. Because the source and object code were difficult to read and understand, it was necessary to attempt to use various documentation aids. Flowcharts, decision tables, and so on were developed and used. However, due to the limited resources, little of this documentation was ever maintained. Moreover, the designers were sometimes not programmers; thus, when the programmers got the design, they would throw it out and start over again (good design, but impractical).

Third, structured methods were in turn developed to try to improve efficiency, flexibility, and communications among the team of developers. Few of these methods actually worked. Surveys show that of literally hundreds of methods developed, few or none are in business use today. Why, especially because some actually were beneficial? They were too hard to enforce and validate. In one case of success for a programming method, the manager took all the code home over the weekend and reviewed it. With this level of oversight, it is no wonder that the staff followed the method. What is the lesson learned here? The lesson is that a tool must be capable of being monitored and measured in use. Otherwise, how can you determine its benefits or impact?

A fourth problem was business unit involvement. Previously, there was little for business people to do on the project. They could not understand the code or do much testing due to its complexity. When implemented, the system or technology would work side by side with the business process. Because the system was so unwieldly, there was no real hope of merging or melding these together. The classic movie example of this was "Desk Set" with Spencer Tracy and Kathryn Hepburn, where departments were to be computerized and made efficient through batch processing. It was a classic failure.

This limited discussion provides the background of why people behave in the old ways and why belief remains strong in traditional methods. You know that things have changed in terms of tools, methods, the power of the technology, and attitudes of business. Yet, there are still old patterns and habits that are difficult to break. After all, people still teach younger students the old methods.

However, not all of this can be thrown out. There is a need for an organized approach. To an extent, it has to be somewhat sequential. But you will probably agree that an overall approach should reflect modern conditions.

WHERE ARE THE BENEFITS?

Forget about the intangible benefits. So what if it looks better or is simpler? These qualities only become benefits if they can be translated into tangible savings, revenue, or productivity. Thus, ease of use is translated into reduced training. Table 1.1 provides a list of tangible benefits and suggests how they can be measured. Table 1.2 provides a list of intangible benefits and how they might be converted into tangible benefits. It is okay to mention the fuzzy benefits as an afterthought but not as a central theme.

Why take such a hard line? Because you have to be able to cull out the requests for new systems and prioritize future work in terms of its impact. If you equate tangible and intangible benefits, you risk credibility for IT and jeopardize future management support.

Table 1.1

Examples of Tangible Benefits

Benefits	Comments
Ease of use	Less training, fewer errors
More of the process covered	Fewer shadow systems and manual work
Simpler transactions	Faster to do the work, productivity
Productivity higher	Reduced time per transaction
Simpler work	Restructured jobs and descriptions
Reduced staff required	Reduced cost of work
Reduced errors and rework	Higher volume of work/time (more throughput)
Faster processing of work	Higher volume of work/time

Table 1.2
Examples of Intangible Benefits

Benefits	Comments
Simpler	Ease of usee
Less complex transactions	Lower labor cost per transaction
Friendlier system	Less trained and experienced staff needed
Better fit with process	Productivity higher
Easier to support	Lower support costs

Everyone talks about benefits when people approve an implementation project. Yet it is interesting that this is forgotten after implementation has been completed. Is it because people fear that there will be no benefits? Is it because the benefits at the start were overstated? Or, is it because they did not measure the business process before they began the implementation? Whatever the reason, it is not uncommon. How should you address this? Following are some guidelines:

- Measure the business process that will be supported by the new technology when implemented. This will provide a baseline for later comparison.
- Measure at the general process or department level, as well as at the detailed transaction level. This will provide validation for general measurements.
- Measure the improved process and new system at the transaction level and work your way up to the process. This will provide a basis for before and after comparison.

A REAL WORLD APPROACH

This discussion can now be placed in a framework that serves as the basis for this book. The approach is based on the following:

- **Importance of business processes in implementation.**
  IT has limited resources overall. There are precious few when you deduct the overhead for ongoing support of current technology. Therefore, IT must focus its energy on those business processes that are critical and where the tangible, strategic benefits are the greatest. At every stage of decision making in implementation, the business process must play a role.
- **Continuous involvement by business unit managers and staff in implementation**.
  Hand in hand with the business process are the roles of the business unit

managers and staff. There is no room for spectators. Business people must be players. They must be involved in more than one half of the implementation activities. Through participation they can shape the system, and through understanding they can define the details of the business process that will fit best with the technology. Implementation is mainly a game for two players—IT and the business. Fortunately, the modern tools and methods support this by enabling many activities to be performed by business staff.

Some of you, upon reading this, might say, "Good idea, but these people are too busy," or, "They are not interested." Well, do you want to implement something for someone who is not interested? If they are too busy to determine their own destiny through participation, are they going to use the new system after implementation? Probably, yes, but not with recurring energy.

- **Comprehensive approach to implementation that includes politics, organization, and process as well as systems technology and procedures.**
  If you ignore the organizational or political factors, you are much more likely to fail. You may suffer *midinstallation paralysis* in which resistance results in inertia. Specific guidelines for dealing with these factors are presented in each chapter. Note that this is not a book about total change and upheaval (one definition of reengineering); rather, it is the implementation of a changed business process that totally integrates with the technology. After success in implementation and process change, there can be organization restructuring and change.
- **The need to implement technology in synchronization to the new or improved business process.**
  You cannot design and implement a new system without regard to the business process it is intended to support. The business process must be defined in part in advance of the technology implementation and then in parallel with it. It cannot be defined ad hoc at the end. You do not want to fit the process around the technology.

ORGANIZATION OF THE BOOK

The approach here (Figure 1.1) is based on creating a process, plan for the business process, as well as a process strategy for implementing the plan (see Chapter 2). Without a process plan, there tends to be a lack of agreement about the purpose and vision of the technology improvement. The strategy provides direction and a roadmap for implementation.

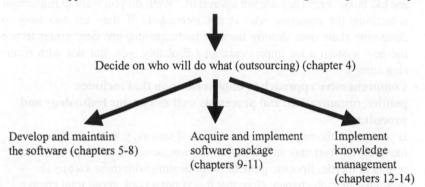

Develop a strategy and plan for the business process (Chapter 2)

Define the new business process along with requirements and benefits (Chapter 3)

Decide on approach and whether the work involves knowledge management (Chapter 3)

Decide on who will do what (outsourcing) (chapter 4)

Develop and maintain
the software (chapters 5-8)

Acquire and implement
software package
(chapters 9-11)

Implement
knowledge
management
(chapters 12-14)

Figure 1.1 Workflow of technology implementation.

With the plan and strategy, you must now determine what the new or modified business process will look like (see Chapter 3). This will determine the general technology requirements. In this chapter, you also decide whether to build or buy and whether the application fits knowledge management. Who will do the work? Will you employ outside firms? This is addressed in Chapter 4. These four chapters get you started and constitute Part I of the book.

The next three parts of the book deal with software development, implementations, maintenance, and enhancement (Part II, Chapters 5–8), software package implementation (Part III, Chapters 9–11), and knowledge management (Part IV, Chapters 12–14). Chapter 15 provides a summary of how the approach can address major reasons for failure.

Many organizations now consider buying packages as a first choice. They do not want to go through the agony and uncertainty of development for routine business functions. The three chapters in Part III address evaluation, selection and negotiation, implementation planning, and implementation of packages. It is useful to point out that many package implementations fail as bad as or worse than development. Thus, guidelines are given to avoid many of the pitfalls.

With expanded computing and communications power and more integrated business processes, more data are available for analysis in data warehouses and data marts. In addition, many organizations are striving to capture best practices in the form of lessons learned databases. Part IV is concerned with knowledge management. Chapter 12 addresses identifying opportunities and a strategy, whereas Chapter 13 discusses services and technology that support knowledge management. Both of these chapters lead to Chapter 14, which addresses implementation of knowledge management.

If this makes sense, then how is it different from other books and approaches? First, it applies more to the real business world. The focus is on effectively implementing the systems and technology in the real world. Many books take a more academic approach that deals with applying methods in a more abstract setting. However, these methods get chewed up fast in the real political and organizational world.

Second, the approach does not depend on a specific method or tool. As you will see, certain methods are favored over others because of effectiveness, measurement, or control. Fortunately, the trend in tool improvement is geared toward a positive management impact.

A third difference is that this approach is more ambitious and comprehensive. Do you understand that in the previous approach if you succeed in implementing a system but fail to improve the business process, you fail. This is radically different from traditional thinking in which an IT person would say, "Well, getting the benefits after the system is implemented is up to the user. It's not my job. I don't control it." Another variation is, "I don't work in the business so how can I change it?" To be successful, IT must establish partnerships with the business to undertake implementations. If business units do not want to participate and change, then it is better to know this at the start and move to another business process. Shoving a system down someone's throat is just asking for trouble.

Each chapter (with the exception of Chapter 15) is organized in a similar manner. This organization is as follows:

- Introduction — provides background to the subject
- Core material of the subject — includes specific steps; identifies and addresses specific problems
- Reduced schedule and cost approach — shows how to implement with limited time and resources
- Examples — provides examples, that occur in the real world.
- Lessons learned — provides tips and guidelines based on experience
- Summary — presents highlights of the chapter and impacts of the materials
- What to do next — illustrates actions you can take to use the material in the chapter

TRANSITION TO A NEW APPROACH

The approach that is presented in this book may be very different from your current approach. If you attempt to radically modify the work priorities and methods, you will likely run into resistance and increase your chance of failure. Costs and schedule will both increase. To head this off, select one area for a test of the new methods. For example, pick one business process and then implement both process change and new technology. With a successful track record, you will then have an easier time expanding the approach. This method also allows you to collect experiences and lessons learned that will give you and others more confidence in implementation.

EXAMPLES

Five examples from different industries are used in this book. Atlas Bank is a regional bank that embarked on major system development. Vision Insurance is a health insurance company that implemented a major software package. Millenium Manufacturing is an Asian manufacturing firm that was faced with modernizing in the face of increasing financial pressure in Asia. Secour Retailing is a major drug store and clothing retail chain with more than 700 locations that implemented software systems as well as knowledge management systems. Roberts Agency is a regional transportation agency that implemented intranet/Internet software as well as software packages. Each of these fictitious companies is based on a combination of real world organizations.

Also included are one-time examples that pertain to a specific topic in the chapter. For example, an IT group in an oil company developed an oil line accounting system for a subsidiary mining firm. The group was very isolated. The parent company sold the subsidiary to a competing firm. Yet, no one told the IT group. Incredibly, they continued with development. When completed, they contacted the subsidiary. Only then did they learn about the sale. Truth is stranger than fiction. This example highlights in an extreme way that business unit–IT communications must be continuously maintained.

LESSONS LEARNED

- **The role of politics—you cannot escape it.**
  This section contains suggestions from experience related to the material covered in the chapter. A basic lesson learned here is that politics in the organization and with suppliers are an integral part of system implementation.

You can try to make technology implementation theoretical and to reduce it to a checklist or a plan. This will help, but when all is said and done, system and technology implementation are very political by nature. Implementing a new system has an impact on the business process. Changing the business process alters the organization, which in turn affects the roles and relationships of individuals and departments.

You should now ask, "So what?" Well, this means that each chapter has to include a political component. How will you deal with the political issues involved in the specific step?

- **Reengineering often fails unless it is linked to implementing new technology.**
Often, trying to reengineer while keeping systems and technology the same only results in downsizing and layoffs. As you have observed, this often means that the good people leave and the "turkeys" stay. Why does reengineering depend on technology? Because you are limited in what you can do with manual forms, procedures, and policies. With technology, you can track workflow and procedures. In the same way, it makes almost no sense to implement a new system without changing the business process. You will spend a great deal of money and derive little benefit.

SUMMARY

In the late 1960s, technology was so limited and complex that any business factors were dismissed. One of the authors remembers telling business staff that, although their requests were valid and had merit, they could not be done. It was the same as the discussion on requirements earlier. The business requirements had to be kept to a minimum as programmers concentrated on specific transactions and file processing. This instilled a general attitude among many IS and IT professionals of distance from the business. Fortunately, new technology and the importance of systems to the business have made the link between business processes and the supporting technology critical.

As seen in this chapter, there are a number of consequences of this interdependence. First, you must be selective in what you implement in terms of which critical processes are affected. Second, implementation counts and gives benefits only if there is a positive business impact. Third, it does not make sense to implement a new technology or system and leave everything else the same. This leads to ensuring that business unit staff and managers are heavily involved in implementation of both the technology and the process change that fits with the technology.

WHAT TO DO NEXT

This section provides some recommendations for you to take after you read the chapter. A basic task is for you to take stock of what is going on in terms of implementation today. Here is a list of questions you should address.

- What is the current division of resources in terms of production support of software applications, maintenance, enhancement, new development, emergency fixes, and implementing packages? Have there been any changes in the last 2 years?
- How are new ideas for systems and technology originated and proposed? What is the review procedure for approval? How are benefits determined? Are subjective, intangible benefits, as well as tangible benefits, allowed?
- Related to the previous questions, how are maintenance and enhancement requests processed and prioritized? How are benefits, defined for these changes?
- Is a backlog of implementation requests maintained? Is the backlog ever reviewed? What percentage of the items in the backlog are ever worked on? What did the business units do in cases where the request was never addressed?
- Is any serious effort given to finding solutions to requests that are not addressed? Or are business units on their own?
- Have the methods and tools employed in systems implementation and support been inventoried?

Chapter 2

Develop the Business Process Plan and Strategy

INTRODUCTION

There are two parts of defining the new business process and business plan prior to making a decision on building or buying software. The first part is to define an overall strategic plan for the business process (*process plan*) and an implementation strategy (*process strategy*) that provides a roadmap for implementing the plan without disrupting the business. This part constitutes this chapter.

Chapter 3 follows up on the strategy and plan by defining in detail the new process with its requirements and benefits. Requirements can be divided into process requirements and technology and system requirements. Traditionally, user requirements fit within process requirements. Both requirements and benefits are grounded in detailed analysis and comparison of the current business process with the future one. This information can be employed to update the process plan and process strategy.

Why go through all of this? Why not just proceed the old way and gather requirements? That is, why not get a pad of paper and conduct interviews? This approach fails for several reasons:

- To understand future requirements, you have to understand how the current process works.
- Gathering requirements will only give you a partial picture based on what people are aware of when you collect the information. Such information is incomplete and may be biased based on their feelings and on what they remember at the time.
- You have to observe the business process to ferret out the shadow systems and determine how exception transactions are addressed.

19

OBSERVATIONS ON BUSINESS PROCESSES

We begin with a look at business processes. A *business process* consists of activities that handle the processing of transactions or work and produce specific results. Business processes can be informal or casual in that they lack formal procedures or systems, or they can be very formal and structured. At one extreme, they can be highly structured such as automated teller machine (ATM) transactions. At the other extreme is strategic business planning that involves creative thought. A business process often starts casually when a company is getting started. As time goes by, the process changes to fit the situation. It has to evolve to take on new types of work. Eventually, the process is shaped by many factors, some of which are as follows:

- *People who work on the process.* Each new employee who works on the process brings his or her skills and experience (or lack thereof) to the process. That is, an employee's previous practices have an impact on his or her current work.
- *Training and procedures for staff.* If there are formal materials and training that are implemented, then each new person learns the process the same way. If, as in many cases, it is informal and learning occurs on the job, then each person will learn differently and work will be done with greater variation.
- *Changes in workload and types.* If a new type of work or transaction has to be addressed, can the process handle it? If not, then what do people do? Often, they will create a manual exception process. As these exceptions increase, people may try to make them more efficient by creating a shadow system. An example occurs when business staff requests changes for an online system from an information technology (IT) group. Suppose they are told that it will take too long. Desperate for a solution, they hire a student intern to program something in a spreadsheet. Voila! A new shadow system has been created. The longer this continues, the more dependent the department is on the shadow systems, and the less real work is performed by the basic system. Thus, if you just replace the basic system, you miss the shadow systems where more than one half of the work is being done.
- *Management turnover.* Each manager may want to put his or her stamp on the process. They may do this by organizing work differently or by creating more bureaucracy and reports. There is cumulative overhead.
- *Other departments and processes.* If one department that feeds work into your process changes their process, you will likely have to modify yours to accommodate this new input.

Each of these factors can have both positive and negative effects on the process. In many cases, if work is not performed to tune or improve the process, gradual deterioration takes place. What types of deterioration can occur? Here are some examples:

- The process becomes so specialized that only certain people can perform certain transactions. If a person is sick, all transactions handled by that person wait until his or her return. Knowledge is spread across many people, and the process loses any economies of scale.
- The process has little central process. Every piece of work is classified as an exception type. This is a nightmare.
- There is a lack of management will. The system appears impossible to change; however, changing organization and staff are impossible options. Thus, the process stays the same and things get worse.
- As workload and exceptions increase, the extent of manual labor increases and the value of the procedures and training decrease.

Organizations knowingly can allow deterioration to occur. In a large bank, management wanted to issue a new credit card. The IT group could not respond fast enough, so management decided to set up a manual group, contrary to the standard process. Eventually, the card only attracted 500 customers. This was too small of a response to be automated but too large to be eliminated. Another manual empire was created and sustained. Staffing grew due to the deterioration of a manual process.

Why is the deterioration allowed to continue? It happens gradually. Many managers are not attracted to these everyday events because there is no short-term fix and because it does not appear exciting. So the process continues. Radical reengineering can be attempted to identify and address these processes, but many of these projects fail. Why? A major reason was that the reengineering did not consider the technology and how it could integrate with the business process. Reengineering also missed the shadow processes.

COMMON PROBLEMS WITH PROCESSES

Although this is not a book about business processes, you should be aware of some common problems with processes so that you can spot them when you perform analysis. Following are 11 problems that have been encountered repeatedly:

1. The process is chopped like a salad. Many different people handle a transaction; thus efficiency is greatly reduced. You cannot find a specific piece of information easily.
2. The current system is only used as an afterthought at the end of the process. All the preceding work is manual and labor intensive. The current system is of little value to the group.

3. The staff employs workarounds to the system. Data elements in the system are used for purposes other than those originally intended. For example, a fax number might be entered for a person's middle name.

4. There are no formal procedures. That is, the process is not documented. You can detect this if you see a bunch of yellow "sticky notes" hanging on the office walls or in cubicles. Everyone has their own way of using the system.

5. There is one person involved in the process who knows everyone and is the base of knowledge about the process. If this person is sick, the process takes a dive. Beware of this person. If you modernize the process and system, their empire disappears.

6. Work is being performed in shadow systems. Data is entered twice — once in the shadow PC systems and once in the regular system.

7. There is hidden rework. People routinely have to correct errors so that most transactions are handled several times. Efficiency is lost. However, because there is no automated tracking of this, there is no management awareness of the extent of the problem.

8. People make manual log entries for each transaction as it passes their desk. This is common in departments where there is fear of intimidation. The motto is "Keep a record of everything so that you have your XXXX covered."

9. There is no measurement of the business process in terms of costs, errors, throughput, processing time, and so on. No one really knows what is going on.

10. Employees do not care. They have stopped requesting improvements because past ideas have been met with indifference. How do you detect this? Ask people what their oddest transaction was. If they look at you with a blank stare, then you can sense a lack of interest.

11. Interfaces among departments are poor. Process work is split among several departments. One department does its part. The output then just sits on a desk or table for hours. So much for overall processing efficiency. No one is in control of the overall process.

View problems such as these as opportunities for improvement. If there were no problems, then you probably would not consider implementing a new system.

STEPS IN DEVELOPING THE PROCESS PLAN AND STRATEGY

Following are the steps for assessing the process and developing the plan and strategy:

- *Step 1: Analyze the current process and develop a process score card.* This gives you a good idea of the existing issues and opportunities that a new process and system can address.
- *Step 2: Develop a general concept about the new process.* The concept outlines what roles of organizations and types of automation are appropriate.
- *Step 3: Develop the process plan.* The plan addresses how you will progress in concept from the situation in Step 1 to that in Step 2.
- *Step 4: Define the business process strategy.* This defines how the plan is to be achieved.

You could spend months analyzing every transaction in the process in great detail. This would be a return to the 1920s industrial engineering. You should instead cover only a sample of transactions and only 5 to 10 steps per transaction. Gather information by being trained to do the process. Perform the process yourself if you can, given health and safety considerations. As is emphasized in Chapter 3, avoid interviews. They are too formal and rigid.

What are the benefits of this work? First, you will have an idea of the potential for improvement. Second, you can identify and eliminate those processes for which there is little room for improvement or for which technology does not appear to have great potential. Third, you can raise the level of interest among managers and employees in fixing the process in addition to the system. Fourth, you pave the way for estimating benefits and gathering requirements. You are likely to develop substantially more process plans than you will initially handle for project work.

It seems obvious that you would not want to embark on getting a new system without a process plan. Yet, people often approach it this way. Why? They think inside the systems box. They do not consider the process to be part of their puzzle. This is wrong.

STEP 1: ANALYZE THE CURRENT PROCESS AND DEVELOP A PROCESS SCORE CARD

One of the first things you want to do when you look at a business process is to measure it. You must know from where you are starting. Go out and observe the process. This is ground zero—where the benefits will be. Identify the specific individual tasks in a representative set of a few transactions. Pinpoint the issues and problems with the current process using direct observation and people's comments. Avoid formal interviews as they tend to elicit canned responses.

A technique drawn from experience is to create a process score card or report card. This can become an annual event. You can take 5 or 10 processes and issue report cards on how each person did. Then you can track this each year to detect trends in deterioration.

What is in the score card? The score card cannot be subjective or it will not be credible. Following are some measurements used in the past:

- Total number of people who spent more than 25% of their time involved in the process
- Turnover of staff in the process per year
- Extent of formal training in the system and process for new employees
- Absenteeism and illness absence days
- Total cost of labor for the process
- Total volume of work (transactions) processed
- Average time of processing transactions based on a sample
- Age of system supporting the process
- Cost of systems support
- Ratio of systems-related costs to total cost of labor and systems
- Number of identified exception types
- Estimated volume of work that is classified as exceptional
- Extent of measurable rework

From these factors you can compare different processes. This score card is also part of the process plan because it gives you insight into the current situation. Consider a series of comparative graphs for different processes using the score card. After you do this the first time, you are likely to be surprised in that much management attention and IT effort are probably being devoted to the wrong processes.

At Vision Insurance, there was an old process for opening and processing incoming mail. Mail was manually opened and sorted, and then delivered to data entry. Data entry would then input the information into an old, legacy system. Not only was data quality and productivity low, but also mail was often misplaced or lost. The score card reasembled the following:

Element	Value	Comment
Number involved in mail handling	5	Work only one shift
Number involved in data entry	45	Wide variety of skills
Productivity in data entry	10–35 per hour	Low to medium productivity due to the fact that people must know all types of letters; wide range of performance
Lost mail	150 known missing items	Many more items not accounted for
Total volume of work	1100 per day on average	Volume varies across the week; staffing is constant

STEP 2: DEVELOP A GENERAL CONCEPT ABOUT THE NEW PROCESS

Based on the analysis of the specific transactions, you can begin to define how a new process could improve these transactions with enhanced technology. Here are some guidelines:

- Processing of a transaction should be supported by workflow tracking, which allows for status codes for each transaction step . After each step, the transaction is moved to an electronic queue for the next step. This allows for high-priority work to be rated first. It also allows for the identification of where specific work is in the process and supports productivity statistics for employees in that the system knows how long each transaction remains at each step.
- To the extent possible, all transaction steps should be supported by the system. Leaving intermediate manual steps eliminates much of the benefit of the technology.
- The interfaces among departments for a transaction have to be automated; otherwise, reports from one system are used as manual input into the next system. Automating interfaces means that different departments may have to share the same system—a political implementation issue.
- Paper should be eliminated early in the transaction and at each step. Paper handling slows a transaction to a crawl. This means capturing data electronically as early in the process as possible.
- Consider having customers or suppliers perform data entry of the initial transaction, as well as automate data capture through bar coding, scanning, and optical character recognition (OCR).

Although you cannot cover every transaction, you now have an idea of what the general system will have to do. More important, if you explain this to managers or business staff, it will be easy because you do not have to dazzle them with technical terms. You can put the old way of doing it next to the new way. Do not go into details as to how it would be done. This will sharpen the interest and support for change and show that the process must be changed.

At Vision Insurance, it was clear that there was no workflow tracking of any work. Actual statistics were rough manual estimates plus some data from the old legacy system. A candidate for the new process might employ scanning of mail and then indexing the mail to identify the customers. Even this could be shortcut by using OCR. After indexing, the documents could be transmitted electronically to data entry where final entry would occur. In indexing the documents, the employee would enter the transaction type so that difficult transactions could be routed and handled by experienced staff.

STEP 3: DEVELOP THE PROCESS PLAN

What is in the process plan? Here is an outline of what is to be included; it can be either a document or a presentation:

- *Introduction.* This section gives the importance of the process plan and what it is to accomplish. The purpose of the plan is to provide a blueprint for the long-term future of the process. The benefits of the plan are to give direction and gain widespread acceptance.
- *Analysis of the current process.* The score card is presented and discussed, along with issues and problems with the process (not just the system). This section now provides specific focus to the process.
- *Examples of current and future transactions.* This section provides a side-by-side comparison of key transactions. There is no discussion of what technology will be needed. In fact, the technology may not even be available yet. This now shows the level of detail that you considered. It also demonstrates that change is feasible and has benefits.
- *Overall view of the new process.* Now step up to a higher level. Provide an overall view of the new process that shows the its simplicity.

What does this presentation outline resemble? It resembles many television commercials that work. They gain people's interest and make them want to move ahead. This is the lead-in to the process implementation strategy. You may want to wait until the strategy is completed before you give a presentation. This would allow for a more complete picture. However, if you split it up, then you will capture people's interest and get more involvement and participation in defining the strategy.

At Vision Insurance, the process plan focused on homeowner's insurance because it had the highest volume and the widest variety of transactions. The side-by-side comparison was based on the approach using scanning. The benefits included productivity and workflow statistics, workflow tracking to identify any transaction, reduced labor in handling paper after the mailroom operation was completed, and improved data entry productivity because the work could be specialized by the skill level of the individual employee.

STEP 4: DEFINE THE BUSINESS PROCESS STRATEGY

The process plan is nice, but what is next? What should you do? You have to make recommendations in your implementation strategy for the process for each of the following factors:

- *Process.* To what extent will the process and its procedures be changed? How will policies be changed?
- *Systems and technology.* What general approach will you use? Will you employ packages, developed software, or both?
- *Roles.* Who will carry out the change and implementation? Will you outsource? Will training, management, and supervision be changed?
- *Organization.* To what extent can the organization be changed? Will the job descriptions of the staff be affected?

The implementation strategy is the sequencing of the changes and focus in each of these areas. Following are some examples to make this tangible:

- Example 1: Moderate approach
  - Stage 1 — Improve the process with the current system. Begin to implement the new system. Involve users heavily in implementation. Do not touch the organization.
  - Stage 2 — Implement the system. Change the process to mesh with the system. Business staff take over many functions and support. Minor changes in job definition occur.
  - Stage 3 — Implement organization changes after the process is stabilized.
- Example 2: Radical reengineering approach
  - Stage 1 — Dramatically restructure the organization and process to get rapid results (often downsizing). Plan to implement a new system after changes. Use consultants widely to perform most of the analysis. Consider outsourcing the entire process.
  - Stage 2 — Begin the implementation of a new system that fits the changed process.
  - Stage 3 — Complete the implementation of the system. Revisit the process through radical reengineering later.
- Example 3: Traditional systems implementation
  - Stage 1 — Analyze and gather requirements for a new system. Leave the current process alone. Limit involvement of business unit staff to interviews and providing data in structured interviews.
  - Stage 2 — Implement the system potentially involving consultants in implementation only.
  - Stage 3 — Potentially make changes to the business process after the department has used the system for some time.

Obviously, there are many more potential strategies. These three examples are often the most typical. The traditional life cycle approach is the third example. Reengineering where you start with a "clean slate" is the second. The first is the one that has been most successful in many projects in different industries and will be the general one employed here.

For Vision Insurance, the near term strategy was to spread staff across two shifts so that the workload would be more evenly distributed. The second part of implementation was to install a scanning system for data capture. The third part of the strategy was to implement a client-server system that linked to the scanning hardware.

HOW TO BUILD YOUR FIRST PROCESS PLAN

Start with a process that crosses two departments. Also, make sure that the process is not politically active or embroiled in disputes. The key search terms are "mundane" and "medium size". The first time out is a learning experience. You will have to do much of the work yourself because most people require a trigger or model to demonstrate what they are after. This means that you should develop a model, scenario, or "straw man" of the process plan and strategy. These trigger reactions from people and get them involved. Do not just sit there and ask them what they think. Many people are so involved in their daily work that they have not had time to consider changes or improvement to any extent. However, they can react.

After you collect information and build your first "straw-man" plan, seek reaction. Show people how their input changes the plan. Get them involved in defining benefits and issues. Make sure that you maintain almost continuous contact.

WHAT CAN GO WRONG?

At this early stage in implementation, you would think that not much can go wrong. After all, you have not spent much money. You have had little opportunity to make any enemies. What can go wrong?

• *Resistance to change.* People see where the plan and strategy are going. They see that they will lose political power or organizational clout. They think it is easier to shut it down now rather than to wait until the implementation steamroller gets going. In some cases, they are correct. Where can resistance arise? We examine several sources and suggest some countermeasures. Business unit staff may be threatened by change. The best approach is to involve them in the plan development. This will increase their confidence. Another source is middle-level business management. This is one of the most difficult to deal with because many changes will restructure their jobs and work more than those of the staff. You might try minimizing the impact and stressing the details of change. Upper-level management may have approved similar projects in the past that failed. Indicate and stress the low risk and confidence in change.

• *The process selected for this is too large or complex.* You cannot get your arms around the process plan. Nothing seems to work. People have a hard time supporting something that is so large. What should you do? Segment the process within departments and work on parts of the process. This bottom-up approach will gain more support. How do you guard against getting less than dramatic results since you watered the process down to the department level? After performing the transactions at the department level, consider the overall process across departments to show how it can be improved. This will also allow you to indicate problems involved in interfaces.

• *Management fear.* Company management become fearful and embarrassed because of the problems now visible with the process. How could this have happened? Who let it happen? What can be done? They may wonder what other problems exist. Here you must provde some perspective that this happens with all processes over time. They need work and improvement.

• *Impatient for results.* The process plan and strategy appear to be okay. However, managers want results faster. This will take too long. That is precisely why some managers go for radical reengineering—to get a quick hit. You should anticipate this and head it off by considering it an alternative and rejecting it because of the potential negative impacts. That is, good people will leave; turkeys will stay. Without technology change, there is only so much that can be done with the process.

REDUCED SCHEDULE AND LOW-COST APPROACH

Do a lot of the work yourself up front. This will help to guide the structure and completeness of the plan and strategy. It will also save time and money. Increase the involvement of others in review and refinement. Develop successive drafts of the plan and strategy in the forms of lists and bulleted items. Design your transaction diagrams as quickly as possible because you want to get a reaction and have people get comfortable with the detail. Use presentations instead of documents. They are simpler and faster to create and modify.

To reduce any gap in time between the plan and strategy and between the strategy and detailed work, you should concentrate on the strategy as soon as you have defined your first version of the plan. Then you can work in parallel on both. This will also give people a more complete picture of what is going on.

Another idea to reduce time is to market the results to managers as you go. In that way, when the final products are complete, there will be no surprises. You will be able to directly proceed to the next step in Chapter 3. We prefer informal presentations in which you walk through the transactions first and then expand management's understanding of the general process from the bottom up.

EXAMPLES

The method presented in this chapter was applied to Millenium Manufacturing, an Asian manufacturing firm. The firm had grown during good Asian times. Seeing clouds on the horizon, they wanted to get new, more efficient systems and processes. They had made one attempt at getting a package, but this fell apart because the package appeared to have few benefits.

The approach was applied to 10 critical manufacturing and distribution processes at Millenium. Here, the sales process is used as an example. Sales crossed marketing, accounting, order entry, warehousing, and billing. Four transactions were identified. One was the most common. A second applied to big ticket items. The other two applied to exceptions. The diagram of a transaction in the current process appears in Figure 2.1. The one for the new transaction appears in Figure 2.2. Note that the new process involves electronic commerce and is quite different from the current process. An intermediate process was identified in Figure 2.3. The evolution is to proceed through Figure 2.3 and toward Figure 2.2 from Figure 2.1. This model was the same for all transactions.

The next step was to show how the process, organization, systems, and roles would change as work progressed (Table 2.1). In this table, the process is simplified in preparation for later implementation. Management fears of risk and positive/negative impacts are eased through the organization and roles entries.

Atlas Bank had spent a lot of money on new software, and the staff claimed to be happy and to think it was a success. Management saw little results from the money. Productivity and staffing levels were the same. Customer service was un-

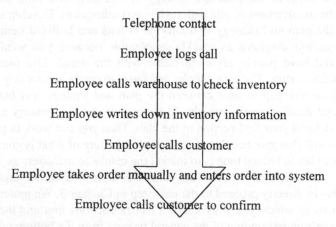

Telephone contact

Employee logs call

Employee calls warehouse to check inventory

Employee writes down inventory information

Employee calls customer

Employee takes order manually and enters order into system

Employee calls customer to confirm

Figure 2.1   Current order handling for Millenium Manufacturing.

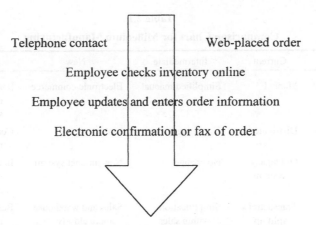

Telephone contact   Web-placed order

Employee checks inventory online

Employee updates and enters order information

Electronic confirmation or fax of order

Figure 2.2   Future workflow for Millenium Manufacturing.

changed. As a result, management was frustrated. At this point, they decided to invoke a policy that all major systems expenditures had to be accompanied by detailed, enforceable savings and benefits. It became readily apparent that the business process would have to be changed to achieve this policy.

The first area of the bank to follow the policy was the credit card and lending area. A process plan was developed for installment loan collections. The benefits were great if changes were made in both the systems and the processes. However, this was only a small part of the business unit. Parallel process plans were developed for credit card, real estate, and leasing in the business areas of application processing, servicing, collections, and charge-off/recovery. With these process plans in place, an implementation strategy was devised.

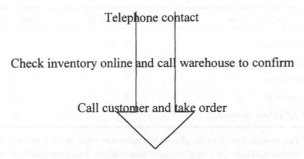

Telephone contact

Check inventory online and call warehouse to confirm

Call customer and take order

Figure 2.3   Intermediate workflow for Millenium Manufacturing.

Table 2.1

Comparison Chart for Millenium Manufacturing

Element	Current	Intermediate	New	Comments
Process	Manual	Simplified manual	Electronic commerce	Some benefits in intermediate stage
Organization	Distributed	Unchanged	Linked	Could change more later
Systems	Old legacy systems	No change	New intranet system	Intranet system to interface with current systems
Roles	Transactions split up	Simplification within sales	Sales and warehouse more closely aligned	Basic roles unchanged

Although space does not permit the strategy to be reproduced here, Table 2.2 shows the evolution of implementation by phases. Each phase took about 1 year and resulted in the changed systems and processes in the business areas.

LESSONS LEARNED

- In developing the process plan, select some sample transactions to validate the plan and to show specifically what will happen with the new process. Otherwise, the new process appears too theoretical.

Table 2.2

Evolution of Change at Atlas Bank

Activity area	Installment lending[a]	Credit card[a]
Collection	1	2
Charge-off	2	3
Servicing	3	4
Payments	4	5
Application processing	5	5

[a] The numbers refer to the phases of implementation. In each phase, the bank could take on a new business area and use their experience in one area as leverage in the next.

- When defining the new transaction, go to where the work is being performed and simulate it in terms of workflow. This will get people involved and help make the new transaction more tangible.
- In presenting or documenting the individual transactions in the plan, use simple flowcharts that reflect current outcomes and future projections. Then, to show the impact of the plan, illustrate the current outcomes with the steps crossed out using the red universal "no" sign.
- In the plan, you identified specific issues pertaining to the current process. Relate these issues to the individual transactions by showing to which steps they apply. Then show how the issues are addressed in the diagram of the new transaction.
- Seek out business staff participation in the development of the process plan. Have them make part of the presentation related to transactions. They have more credibility than you do because they are close to the action.
- In defining a process implementation strategy, define several alternatives, including the three examples given in this chapter. For each alternative, consider building a table such as the following for each alternative. The actions are listed in the first five columns and the benefits are indicated in the last column.

Stage	Process	Organization	Technology	Roles	Benefits

These alternatives can also be summarized for comparison into another table. The data for each alternative are listed in terms of benefits and risks.

Stage	Alternative	Overall benefits	Overall risks

SUMMARY

After reading this chapter, you may be thinking that there is a great deal of start-up work to perform. Why not just get to the detail? Suppose that you do plunge directly into the details of a process. People will wonder what the project purpose and scope are and will question what you are doing. There is no common vision or support for change. Some people do not want to change the process. Others want a new system, but that is it. Still others want radical change. If you spend the time up front, you will be more likely to gain support.

A more basic question is about you. Do you really want to work on a process that has no direction or plan? This seems pointless without objectives and scope. The plan, strategy, and systems are all interrelated.

WHAT TO DO NEXT

1. Select a small process with which you are familiar. Make two lists—one of issues involving the process and the other involving some of the transactions.
2. Now assume that you could apply any technology to improve the process without regard for cost. Develop a workflow of how transactions in new process would work. Compare these with the corresponding transactions in the current process.
3. Based on the work in the first two exercises, define requirements for the new process and system. Write down the changes that must occur to move from the current process to the new process for the transactions you specified.

Chapter 3

Defining the New Process, Benefits, and Requirements

INTRODUCTION

The process has been selected. A process plan and an implementation strategy are in place, providing direction. Now you can get down to the details. Following are the steps:

- Step 1: Define the new business process in line with the process plan and strategy.
- Step 2: Determine requirements for the new process.
- Step 3: Identify generally costs and benefits of the new process.
- Step 4: Present the results to management for review and approval.
- Step 5: Define the technology approach based on requirements.
- Step 6: Determine whether to buy or to build a system.
- Step 7: Decide who will do the work. Outsourcing is considered here and is the subject of Chapter 4.

Why go through all this work? Why not just start gathering requirements and then move to Step 6? Because if you start with requirements, you will likely be drawn to the problems in the current system and will concentrate on replacing it. You will most likely ignore the problems in the process, the shadow systems, and workarounds for exceptions. It is very possible that any new system built or bought will not fit the process plan or strategy without taking the previous steps because of the lack of knowledge.

Put these together and you can see that, if any benefits existed at all, they would be greatly reduced. In some companies, a new system obtained without going through the steps has made things worse.

The goals of the work here as follows:

- Define a new business process that is complete and that can be effectively supported by systems and technology.
- Determine an implementation approach for the new system that will yield benefits and be capable of being installed within budget and schedule with minimal risk.

What are some of the risks? The technology selected may not be established. The effort to interface the new system to current technologies and systems may be greatly underestimated. You might miss some shadow systems leaving the organization to struggle with these later.

The scope of the work can be viewed in four parts:

- Process—all aspects of the business process as well as interfaces
- Organization—all departments that provide support to the process
- Technology—both current systems and technology indirectly and directly employed in the process
- Dependent processes—with every major process, there are typically lesser supporting processes and activities

Leaving something out here can exact a high price later. If you do not involve a specific department, you risk interface problems and hostility later. Omitting a part of a process or technology may increase the interface effort. It is tempting to exclude some processes to keep the scope of the project limited. You are then going to pay the price later when you are forced to retrofit interfaces between two processes.

STEP 1: DEFINE THE NEW BUSINESS PROCESS

For a detailed step-by-step approach, you might consult a process improvement book (see, for example, Lientz and Rea, 1998). Begin with a detailed analysis of the business process. Your work should include shadow systems and exceptions as well as the "run of the mill" work supported by the current system. Your analysis must be carried out at the transaction level because benefits and requirements build from here. Note that this approach is much more time consuming than sitting down and making a feature list for new software— thereby avoiding the process and focusing on the software. Until you define the new process in detail, you really cannot build a feature or requirement list.

You can define the new process in terms of the dimensions listed as follows:

- *Organization.* Which organization will perform the new process? Will it be the same department, another department, suppliers, customers, or an outsourcing firm? The decision has an impact on the systems and technology to be used. Wal-Mart transferred functions to suppliers as have automobile manufacturers. Banks transferred functions to customers through banking on-line and automated teller machines (ATMs).

- *Capture information.* How, when, and where will information be captured? With a new process, consider capturing the data earlier in the process or at the front end of the process. Perhaps, you can capture information through electronic commerce, Electronic Data Interchange (EDI), or scanning and imaging.
- *Output of process.* What happens to output from the new process? Don't restrict yourself to the current process and system. Be creative. Perhaps, if the new system is based on a database management system or a Fourth-Generation Language (4GL), you can cut down programmed reports. Maybe you can establish an automated interface with later systems. Automated interfaces to following processes reduced data entry and errors by more than 30% at Vision Insurance.
- *Who will do the work?* The first dimension determined the organization. Here you define who will actually perform tasks in the process. If you are thinking of providing a simpler, easier-to-use system, then you can lower the personnel requirements.
- *Paper and tracking.* You should try to eliminate manual tracking and paperwork in the new process. If the new system has workflow tracking when you know where every transaction is, then you can cut out logs and forms used to summarize production and track work.
- *Policies.* Every process is governed by policies of the organization. Policies in turn affect the procedures, workflow, and systems. When defining a new process, consider simplifying the policies. This will reduce your system requirements and, perhaps, speed up implementation. Simplifying policies also leads to eliminating exception transactions. In one bank, the policies were changed so that 15% of the workload was reduced and the number of exceptions reduced by 45%!
- *Location.* Where will the work be performed? Which offices and space will be employed? How will the office space be laid out? In one case, the current office space for data entry consisted of partitioned offices. It was too hard to observe the work being done. This was replaced by open space with the supervisors desk on a raised floor, allowing the supervisor to oversee the work. Moving to a better location for a distribution firm reduced employee absenteeism and improved employee morale.
- *Systems and technology.* What new technologies are appropriate to perform the internal work of the process? Can more of the process be automated through these technologies? Such steps could eliminate some of the shadow systems.

In short, make the new process as simple, straightforward, and stable as possible.

STEP 2: DETERMINE REQUIREMENTS FOR THE NEW PROCESS

Requirements are specifications and features for all aspects of the new process. They do not just apply to software and hardware. Here are categories of requirements:

1. Organization
   * Staffing—number of people, education, and experience; specific skills
   * Management—number of supervisors, skills and knowledge of managers needed
   * Support requirements—roles and responsibilities of specific support staff required
2. Process
   * Policies—rules on what and how the work is to be performed
   * Procedures—specific workflow steps
3. Infrastructure
   * General location
   * Offices and buildings
   * Voice communications
   * General office layout
   * Access control
   * Office equipment and furniture
   * Other building and space requirements
4. Networks
   * Local area networks (LANs)
   * Mobile communications
   * Wide area networks (WANs)
   * Internet and international communications
   * Network interfaces
5. Hardware
   * Servers
   * Workstations
   * Portable computers and mobile equipment
   * Mainframe and minicomputers
6. General software
   * Operating systems
   * Utilities
   * Database management systems

- Fourth-Generation Languages
- Personal computer software
- Groupware
- Electronic mail
- Internet communications (browsers, web servers, firewalls, etc.)
- Scheduling and other collaborative software
7. Application software functions and features
   - Input
   - Business rules
   - Processing requirements
   - Output
   - Interfaces
   - Management control

What is the form of these requirements? A good format is lists. Each list has several columns. The first column is the requirement, the second consists of detailed or example specifications, and the third consists of comments on the requirement. You might comment on the importance of the requirement or provide additional detail in the last column.

As you identify these requirements, what do you do with them? First, you have to ensure that they are really necessary. Evaluate this by constructing several additional tables. Begin at the transaction level with Figure 3.1. Second, build to the workflow level in Figure 3.2. In addition you will have general requirements that cross all transactions. For these use Figure 3.3.

Transaction: _____

Step in transaction	Old process	New process	Requirement

Figure 3.1   Transaction-level requirements.

Transaction	Requirements	Comments

Figure 3.2   Workflow-level requirements.

Requirement	Why needed	Comments

Figure 3.3   General requirements.

These tables will be expanded when you consider costs and benefits in the next step. The information in the tables shows how the new process requires these items. Any later discussion of specific requirements can then be related back to the transactions and the workflow. Defining these requirements helps you to validate the new process.

STEP 3: IDENTIFY COSTS AND BENEFITS

Costs

You must develop as detailed a list of costs as possible. A big mistake people make is the omission of certain critical costs. For this reason, a starting list of cost elements has been provided in Figure 3.4. Where do you obtain the cost

Network
- Cabling
- Network facilities work and preparation
- Routers/hubs/gateways/bridges
- Network operating system
- Network management software
- Network security software
- Internet and remote computer access
- Leased lines
- Additional WAN components
- Additional LAN components
- Mobile communications
- Network testing and diagnostic equipment

Hardware
- File server
- Database server
- Application software server

Figure 3.4   Detailed cost elements for the new process.

- Fax server
- Electronic commerce server
- Firewall server
- Printers
- Scanners
- Mobile hardware
- PCs and workstations
- Test environment hardware
- Backup/recovery hardware
- Specific industry hardware
- UPS hardware and surge supression devices

Software

- Server operating system
- Workstation operating system
- Database management systems
- Fourth-generation language
- Internet software
- PC/workstation software
- Electronic mail software
- Groupware software
- Security software
- Backup/recovery software
- Utilities

Installation support

- Cabling and network support — office level
- WAN labor
- Hardware installation and testing
- System software installation and testing
- Interface support
- Testing support

Figure 3.4 (*Continued*)

information? You have requirements from the previous step. To get prices and estimates of work, there are support groups in your company. You can also surf vendor sites on the web and even contact vendors for general information. Be careful here in that you may get the vendors too excited, and they may interfere with your work.

Once you have determined the number and type of component needed for each requirement, you are in a position to justify these. Create another table as in Figure 3.5. In this figure, the justification for the quantity of the item needed

Requirement	Quantity	Justification	Model/type	Justification

Figure 3.5   Validation of sizing and costs.

appears in the third column and the justification for the model or type of item is in the last column.

BENEFITS

The emphasis in this section is on tangible benefits. People have long been promised many things through automation and change. Actual delivery has often been less than stellar. Given the investment of time and money, and the potential risks and disruption of the current process, there must be substantial benefits that can be measured and verified.

Table 3.1 provides a list of benefits and attempts to suggest how to make these tangible. As you can see, you may need to translate benefits several times to get down to tangible productivity benefits.

How do the benefits relate to the business process? To be credible, you must answer this question at both the transaction and the workflow levels. Construct the tables using the corresponding tables from the previous step as shown in Figures 3.6 to 3.8.

Table 3.1

Benefit	Tangible impact
Ease of use	Reduced training, shorter transaction time, lower skill requirements
Reduced time to perform the process	Productivity, lower staff levels
Reduced error rates	Productivity
Simpler policies	Reduced effort, reduced number of exceptions
Reduced number of exceptions	Productivity
Ability to handle more work	Lower costs per transaction
Improved documentation	Reduced training, fewer errors, greater uniformity of work
Increased uniformity of work	Productivity
Better management information	Better tracking of work, raising levels of production
More information from the process	Better marketing of products, improved costing

Transaction: _____

Step in transaction	Old process	New process	Benefit

Figure 3.6 Transaction-level benefits.

Transaction	Benefits	Comments

Figure 3.7 Workflow-level benefits.

Benefits	How realized	Comments

Figure 3.8 General benefits.

STEP 4: PRESENT THE RESULTS TO MANAGEMENT

Presenting the results to management sounds straightforward, but it is fraught with political peril. You risk alienating staff involved in the process or IT staff who support the old system, even if you gain management support. We divide this step into actions.

ACTION 1: UNDERSTAND THE AUDIENCE

The new process affects management, the line organizations, IT management and staff, and other departments. For each audience, you must anticipate their concerns and address these proactively. Here are some examples from past projects:

- *Line organization involved in the process.* People may fear for their jobs. They may resist new methods and feel that the current process is just fine. They may see the project as more work with little benefit. They also may be concerned that the new process implementation will be too disruptive.

- *Other departments*. They may fear collateral damage during implementation. They may also be concerned about how they will have to change to fit with the new process.
- *IT department*. New systems and technology are not always exciting to some IT people. To them it represents a tremendous burden. They must not only continue to support the existing technology but also learn the new. Ultimately, their knowledge of the technology that is replaced is trashed.
- *Management*. Many managers want benefits with no political issues or disruption. They abhor risk.

Begin with these concerns and make an expanded list for each audience. Next, identify individuals in these organizations who should be involved in the meetings prior to the management presentation. Think about ways to head this off. One method is to stress the gradual nature of the change, whereas another is to indicate that there will be the opportunity for most people to provide input.

ACTION 2: DEFINE YOUR PRESENTATION STRATEGY

How are you going to proceed? Should you just prepare a formal presentation on your own and then make the pitch to management? If you do, your chances of success are slim to none. You must sell the new process bottom up—one person at a time. This takes more time but is essential if you want to be successful.

The following strategy has proven itself. Prepare slides of the following:

- Old process workflow
- New process workflow
- Comparison of old and new process showing savings
- Example transaction in detail to validate the previous general findings
- List of shadow systems and workarounds
- Requirements summary
- Costs and benefits summary

These slides will serve as the core of the presentation later. Continue with the strategy as follows:

- Conduct informal meetings with department staff and managers involved in the process and solicit questions and comments.
- Revise and update the slides and meet with IT.
- Now that you have covered the basics, move out to related departments, including internal audit.
- Prepare the formal presentation.
- Make the presentation and follow up on action items.

By taking this approach, you gain support and commitment incrementally. The chances for misunderstandings will be reduced.

ACTION 3: CONDUCT INFORMAL MEETINGS WITH DEPARTMENTS

Begin informal meetings by showing the transaction detail. This will give the audience a point of reference. Then move up to the workflow. Your purpose initially is have them gain an understanding of the new workflow and then agreement that the new process is both feasible and desirable. Modify the materials based on their comments and return. This indicates that you care about their views and have adapted your presentation to include them. In this second meeting, you are aiming at involvement and commitment to support the management presentation. You really want them to participate in the management presentation and to answer questions.

ACTION 4: CONDUCT INFORMAL MEETINGS WITH IT

You want IT to see the importance of the work. You want to calm their fears about change via a discussion of equipment, software, and implementation. Initial meetings here involve the managers and then move down to the analysts, programmers, and other staff. The order of the review is the same as for business departments, but the focus is on the technology and requirements.

ACTION 5: MEET WITH OTHER DEPARTMENTS

Who else should you approach? With an important process, you should include internal audit. For them, you will present the transaction detail and workflow and discuss controls. Any departments that interface to the process through input, output, or other link should be approached. Here you should show the detail, while focusing on how the interface that impacts them will work. They may appear less than interested but should appreciate that you have taken the time to work with them.

ACTION 6: MEET WITH MANAGERS INDIVIDUALLY

Meeting with managers individually may seem superfluous, but it is not. In the general meeting, you will be unable to address each manager's concerns and questions. In this action, you will approach individual managers with one-on-one meetings. Answer any questions and discuss political implications of the change and improvement to the process.

ACTION 7: PREPARE THE PRESENTATION

The outline of the presentation follows that of the initial work. Here is an outline you can follow:

- Introduction (purpose, scope, what has been done to date)
- Old process workflow
- New process workflow
- Comparison of old and new process showing savings
- Example transaction in detail to validate the previous general findings
- List of shadow systems and workarounds
- Requirements summary
- Costs and benefits summary
- Next actions to be taken

Make sure that your first action does not involve spending a substantial amount of money. That will likely delay approval. Management will probably approve the initial steps knowing that if they do not work out, the project can be halted. Actions should consist of determining the detailed approach in terms "buy versus build" and whether to outsource some of the work.

ACTION 8: CONDUCT THE PRESENTATION AND DO FOLLOWUP

During the presentation respond to any questions directly. Draw upon the business departments and IT as you go. Do not answer for them. Give them an opportunity to field the question and summarize what they say. After the meeting, follow up on the action items right away. Do not let grass grow under your feet.

STEP 5: DETERMINE THE TECHNOLOGY APPROACH AND REFINE COSTS AND BENEFITS

In Step 2, you determined requirements for the new process. In some cases, you can specify the technology at that point if the situation is relatively simple. However, there may be a need for further analysis. The purpose of this step is to define the new systems and technology and then update the costs and benefits for the new process. You will then be in a position to make the decision to acquire a software package or build a system.

The scope of this step includes all the hardware, network, and software components but excludes the application software decision. The application software decision is directly affected by the direction you take here. If you select an environment or *architecture* (technology structure) for which there are no off-the-shelf software packages, you will do development. A guideline here is to keep the architecture open to admit the potential of software packages. You will be refining what you identify here as you get further into implementation.

In Steps 2 and 3, you identified general components and attempted to estimate what was needed. This was prior to the management presentation (Step 4) where you garnered support and the endorsement to proceed. You are now in a position to involve the IT group more directly in determining what specific systems and components are needed. The detailed list in the costing figure (see Figure 3.4) still applies and can be used as a starting point.

ACTION 1: DETERMINE THE ARCHITECTURE TO SUPPORT THE PROCESS

To validate the requirements and costs, you should construct an *architecture* based on these components. An architecture is the structure of hardware, software, and network components and their relationships that support the system and, hence, the new process. The architecture specifies the following:

- What components are needed
- The purpose of the components
- How components relate to each other
- Any supporting hardware, software, and network products needed

ACTION 2: DEFINE THE QUANTITIES NEEDED AND SUPPORT REQUIREMENTS

This action begins with determining the quantities of PCs, software licenses, and so on that are required by the process. Note that in the scope of the chapter, you also must consider related departments and dependent processes. These may also require technology upgrades. Beware that the cost and extent of change and updating necessary can be more than that required by the new process directly. In the case of Millenium Manufacturing an entire new WAN was implemented. Its cost exceeded that of the technology and systems for the process by a factor of two!

As part of this action you must also define support requirements. This consists of the following:

- Personnel to maintain, troubleshoot, and manage the technology
- Operations and maintenance support for the systems software
- Hardware and computer operations support
- Vendor maintenance of hardware and network systems

Support costs are not cheap. For Millenium Manufacturing, these costs totaled 35% of the technology investment required. Leaving this out or omitting some items can result in an unpleasant surprise later.

ACTION 3: UPDATE THE COSTS AND BENEFITS

It is clear that the costs must be updated with the new information. These costs are likely to be higher than the original estimates. Therefore, you should scour the areas affected by the technology to determine if there are additional benefits. For Millenium Manufacturing, such a search revealed the following benefits:

- *Modernized PCs.* These reduced support costs as compared with the obsolete hardware.
- *Wide area network.* The increased capacity network can provide for the sharing of engineering drawings and production data. Prior to then, these were transferred manually, resulting in many production mistakes.
- *Electronic mail and groupware.* These software tools were added on top of the network and provided for a collaborative effort to reduce manufacturing costs.
- *Production statistics.* More production statistics were made available online. This allowed supervisors and managers to more dynamically allocate staff resources, thereby reducing overall costs.

STEP 6: MAKE THE BUY OR BUILD DECISION

You are now in a position to make a fundamental decision. Should you attempt to develop the application software yourself or with others, or should you acquire a software package? Within this question are several others:

- If you develop it yourself, what types of tools will you use? Who will do the work?
- If you acquire a package, what software modifications will be needed? Who will make these changes? What vendor support will be required?

There are a number of factors influencing the answers to these questions. These include the following:

- *Availability of the current IT staff to do development.* The IT staff already is filled up with work. Adding more work means that something has to give.
- *Skills and knowledge of the current IT staff.* If you have selected a technology direction and architecture that are foreign to the staff, you must allow for a learning curve if you do development internally.
- *Availability and skill levels of the vendor.* The vendor includes both the supplier of the software package directly as well as the staff of the underlying database management system or Fourth Generation Language.
- *Availability, skills, and costs of consultants.* These consultants can aid you in implementing a package or in supporting new development.

Due to the importance of this decision, you should proceed discretely.

ACTION 1: SURVEY THE SOFTWARE MARKET AND CONSULTING SUPPORT AVAILABLE

You probably have identified several potential software packages. There are typically several packages known in your industry. In each of the five examples, there were at least two to three software packages. You can also search the web for software. Once you have found a package here are some things to do:

- Go to the web site and download available product information. Most vendors offer a basic software package and then add-on modules that provide additional functions. Consider these as well.
- Contact the vendor for basic product information.
- Search the web and literature for software evaluation articles and comparisons that have already been done .
- Build a list of features of the software using one package as a start and then expanding it as you review other packages (use the feature list in Figure 3.9 as a starting point).
- Determine roughly any areas of the new process that cannot be supported by the process. You will not have sufficient information to do this in detail. Here you are looking for basic functions.
- Search the web and literature for consultants who could provide support for customization, implementation, or both.

A major lesson learned here is that you must select the consultant at the same time you pick a package. Many people select the package first and then the consultant. However, it is a total effort involving both. Moreover, the cost of the consultants can exceed that of the package by 5 to 10 times. You could also select a package and later find that there are no suitable consultants. Then it is back to the drawing board.

Hardware requirements
- Processor
- Operating system
- Memory
- Disk storage
- Other hardware requirements

Network requirements
- Network operating system
- WAN support

Input
- Mobile devices
- Scanning/OCR
- Other input requirements

Support for business rules
- Control tables
- Business rule customization

Output
- Printer
- Fax

- Files
- Other devices

Interface support
- Electronic mail
- Groupware
- Internet
- Electronic Data Interface
- Electronic commerce
- OLE/Active X interface

Workflow routing
Security and access control
Ability to add other features easily
Database management and utilities
required
Available documentation
- User procedures
- Training materials

Installation support

*Note that you should add features specific to your industry to this list.

Figure 3.9   General software package feature list.

ACTION 2: DEVELOP AN INITIAL PROJECT PLAN FOR IMPLEMENTING A SOFTWARE PACKAGE

This may seem like a great deal of work based on limited information. It is; however, to roughly determine an implementation schedule, you must construct a project plan. To get you started a template for this has been included in Figure 3.10. Use this and fill in details based on estimates. Here are some guidelines from experience:

- Allow several months to evaluate and select the software and vendor.
- Allow 1 month to gain management approval due to the expense and political nature of acquisition in many organizations.
- Plan on several weeks to 1 month of contract negotiation.
- The start-up time for the actual work will probably be 1 month or more, because the vendor must schedule the people to be with you.

1000 Identify potential software packages
  1100 Determine range of software packages
  1200 Identify potential packages
2000 Define evaluation method
  2100 Identification of roles in evaluation
  2200 Evaluation methodology
  2300 Evaluation checklists and forms
  2400 Review of evaluation approach
3000 Investigate software packages
  3100 Collect data from literature and web on packages
  3200 Collect data from vendors
  3300 Vendor presentations
4000 Identify software consultants
  4100 Gather information on web
  4200 Perform initial screening of consultants
5000 Investigate software consultants
  5100 Contact consultants and obtain information
  5200 Obtain and check out references
6000 Evaluate/select software and software consultants
  6100 First cut at consultants for each package
  6200 First cut at packages and consultants together
  6300 Prepare request for proposal
  6400 Solicit detailed proposals from vendors
  6500 Evaluate proposals
  6600 Identify leading and follow-up vendors and consultants
7000 Make recommendations to management
  7100 Prepare presentation to management
  7200 Conduct informal reviews with managers
  7300 Make presentation
  7400 Follow up on presentation
8000 Develop implementation strategy
9000 Develop implementation plan
10000 Negotiate with vendors
11000 Implement infrastructure to support the package
  11100 Hardware and system software
  11200 Network
  11300 Testing of technology infrastructure
  11400 Nontechnology infrastructure
12000 Install the software

Figure 3.10 Project template for implementing a software package.

13000 Training and initial use of software package
   13100 Identify staff to be trained
   13200 Determine training and initial use requirements
   13300 Conduct training
   13400 Initial period of use
14000 Convert the data to the new system
   14100 Analyze current data
   14200 Determine conversion approach
   14300 Carry out data conversion
   14400 Review conversion results
15000 Customization of software
   15100 Define areas of customization
   15200 Determine approach to customization
   15300 Define benefits for customization
   15400 Determine contract terms for customization
   15500 Carry out customization
   15600 Documentation
16000 Implementation and testing of customization changes
17000 Interfaces to other systems
   17100 Define interfaces that are needed
   17200 Assess current interfaces and define interface approach
   17300 Determine detailed interface requirements
   17400 Carry out interface design
   17500 Interface programming
   17600 Interface testing
   17700 Documentation
18000 Conversion from old process

Figure 3.10 (*Continued*)

ACTION 3: MAKE SOME PRELIMINARY DECISIONS ON THE DEVELOPMENT APPROACH

Although you have not decided on development, you can determine in general the development tools you will employ and the development environment. These include the following:

- Database management system or Fourth Generation Language to be used (e.g., Oracle, Informix)
- Languages to be employed (e.g., COBOL, C^{++}, Java, Visual Basic)
- Other software development tools (e.g., developer workbenches, program libraries)

You can also consider who will do the development. Are there adequate internal resources available? If not, then you can evaluate several vendors as you did for the software package in Action 1.

Next, you must determine what development approach you will take. One approach is defined in the second part of this book and involves rapid prototyping. You could use a traditional life cycle approach. This decision will provide structure to the project plan in the next action.

ACTION 4: DEVELOP AN INITIAL PROJECT PLAN FOR DEVELOPMENT

With Action 3 out of the way, move to building a project plan to estimate the duration of development. A template for rapid prototyping is given in Figure 3.11.

1000 Determine methods and tools for development and production
    1100 Analysis and design methods and tools
    1200 Programming environment
    1300 Testing tools
    1400 Integration methods and tools
    1500 Interface methods and tools
2000 Acquire tools and establish methods
3000 Training in methods and tools
4000 Develop strategy for development
    4100 Client software
    4200 Server software
    4300 Interfaces
    4400 Business process
5000 Client software
    5100 User interface
    5200 Workflow routing
    5300 Navigation among screens
    5400 Interface with server software
    5500 Determine documentation requirements
6000 Database
    6100 Define database for new system
    6200 Evaluate database with new business process
    6300 Compare database with existing files and databases
    6400 Assess interfaces to other systems
    6500 Determine documentation requirements

Figure 3.11 Project template for developing software using prototyping.

7000 Server
 7100 Definition of business rules
 7200 Design structure for server software
 7300 Identify object and transaction classes
 7400 Determine documentation requirements
8000 Initial prototype
9000 Successive prototypes
10000 Interfaces
11000 User procedures and training materials
12000 Data conversion
  12100 Analyze current data
  12200 Determine conversion approach
  12300 Carry out data conversion
  12400 Review conversion results
13000 Testing of system
14000 Testing of new process
15000 Implementation and cut-over to new process

Figure 3.11 (*Continued*)

ACTION 5: PREPARE COST AND SCHEDULE ESTIMATES FOR EACH ALTERNATIVE AND DETERMINE THE RISKS AND EXPOSURE FOR EACH

You may have several alternatives. In the case of Millenium Manufacturing, there was one internal development alternative, an external development alternative, and two software package alternatives. For each, you will estimate costs. Examples for both development and packages are given in Figure 3.12.

General for both
  New process
  • Definition of detailed transactions for new process
  • Documentation of business rules
  • Development of procedures and policies for the new process
  • Forms development
  • Training materials for new process
  • Training in the new process

Figure 3.12   Cost elements for development and software package alternatives.

Data conversion
- Analysis effort
- Data conversion
- Testing of converted data

Interfaces
- Analysis and design of interfaces
- Programming of interfaces
- Testing and integration of interfaces
- Documentation of interfaces

User installation and cut-over
- Cut-over to new process
- Monitoring of new process

Software package

Initial work
- Analysis/evaluation of software packages
- Contract negotiation with software vendor
- Analysis/evaluation of consultants
- Contract negotiation with consultants

Installation support
- Installation support for the package
- Setup of tables and parameters of the package
- Training of IT staff in the package
- Training of business staff in the package

Software modification
- Modification analysis of the package
- Modifications to the software
- Testing of modifications
- Documentation of modifications

Development

Requirements analysis for prototype
Client software design
Server database design
Client software development
Database setup
Server software development
Unit testing
Integration testing
Documentation for system

Figure 3.12 *(Continued)*

ACTION 6: MAKE THE RECOMMENDATION TO MANAGEMENT

Here you will make the recommendation to management. An outline for the presentation might be as follows:

- Introduction
  - Purpose of the software
  - Summary of the new process and technology
- Approach and actions used in analysis
  - The development alternatives
  - Approach
  - Discussion of alternatives
  - Costs
  - Risks
  - Benefits
- The software package alternatives
  - Approach
  - Discussion of alternatives
  - Costs
  - Risks
  - Benefits
  - Recommendations
  - Action items

WHAT CAN GO WRONG?

From past projects, here is a list of things that can go wrong:

- The software package is selected before the new process is even defined. This will likely be very expensive because there are so many unknowns. The effort to define a new process and proceed with the steps in this chapter may stretch out the time but only entail labor effort. These steps greatly reduce the risk.
- Parties overpromise what they can deliver. IT may promise to do development but lack resources. Business departments may indicate that they will pour resources into the project. Vendors may indicate that their software will meet almost all needs without modification. Part of this stems from people's optimism. Don't accept any of this. Reserve judgment until later.
- Management often attempts to pin down these cost and schedule estimates as the final ones. This is impossible without further analysis. The estimates are necessary to make a decision. Head this one off by stating repeatedly that the estimates will be revised.

REDUCED SCHEDULE AND COST APPROACH

This chapter has defined a great deal of work. If you attempt to do it alone, you will never finish. If you attempt to be too precise, you will fail. The first suggestion is to involve people in the business, IT, and your current vendor staff to help in the analysis. This needs to be a collaborative effort. If they participate, then there is more likelihood that they will support the results and be committed to the project.

A second recommendation to speed things up is to generate outlines and structures for documents and presentations at the beginning of the project. Then flesh these out as you go. A related idea is to market the results in an informal way to managers as you go. This will reduce the time it will take later for them to understand what you have done and subsequently the time to gain approval.

EXAMPLES

Millenium Manufacturing ultimately selected to do their own development. Their decision was based on the fact that even though the new process simplified the current process, there were still unique features that could not be addressed by any existing software package.

Secour Retailing selected a package but did not do the analysis of the first step. They roughly estimated costs and benefits and gained management support. The package was later acquired and then found to be lacking in key features. The result was extensive customization.

Roberts Agency, the transportation agency, performed the steps and concluded that they would have to do both development and package acquisition. There was no package that could handle certain functions; however, if those functions were excluded, there was a package with a close fit.

LESSONS LEARNED

- Concentrate on no more than five sample transactions in defining the new process. The detailed analysis for all transactions will be performed later.
- Identify the shadow systems and workarounds early. Get recognition of these from the business department. This will gain their support for the new process. Give attention to these in presentations to illustrate the shortcomings of the current system.
- Obtain a list of people who can review what you have done early on. This includes a business staff member who can review the process work you do

and the benefits and an IT person who can review architecture, technical approach, schedules, and costs.

SUMMARY

In this chapter, you defined the overall new process along with requirements and benefits. You then proceeded to develop estimates for costs, schedules, and project plans. None of this will be complete. However, these are essential to make the decision on what direction to take.

This chapter has defined major steps that must be taken prior to plunging into development or software acquisition. If necessary, you should spend more time here because every additional hour will likely save much more time later. The steps have been defined in a linear way. However, you should consider starting these in parallel and then building the end products in parallel.

WHAT TO DO NEXT

- For a current project that is underway—either a package implementation or development project—review what has been done and what was done to address the steps in this chapter. You will find that the steps were probably not carried out overall. What impact has this had in terms of the project? What surprises have surfaced?
- Practice carrying out the steps on a very small process within your department. Walk through each step to determine what could be done. Ask the following questions:
  1. What political barriers would you face if you really did the work?
  2. What is the condition of the current process and technology? To what extent do business and IT staff recognize this?

Chapter 4

Assess Outsourcing

INTRODUCTION

Outsourcing occurs when a company contracts with a supplier or vendor to provide specific services and support for multiple information technology (IT) activities. Note that outsourcing typically refers to services and not purchased products. Providing an ongoing service means that there is a need for both a managerial and technical relationship between the supplier and the customer firm. Another observation is that most outsourcing agreements cover several years (up to 5–10 years in some cases). Why this long? Because of the time it takes to come up to speed and become efficient as well as the practical infeasibility of switching suppliers quickly.

Outsourcing of information systems (IS) has been going on much longer than you might imagine. It began in the 1960s with facilities management. Companies had no IT or IS organization and did not know what to do. Therefore, they contracted with firms such as IBM, Computer Sciences Corporation, and others to run their computer center. Firms such as these also provided programmers and systems analysts on-site to do work on a time and materials basis.

Another example was the use of service bureaus. Companies who required data processing but could not afford their own computers used service bureaus. They brought their data to the service bureau who processed it and returned output to the company.

A third example was external timesharing. Computers inside companies in the late 1960s and early 1970s were batch processing machines. To get access to online systems, you had to use a service bureau or timesharing firm. Charges were based on usage. This was extremely expensive, but firms had little choice. Outside timesharing for some firms ate up 10% of their IT budget.

As the 1980s progressed, companies eliminated outside timesharing and service bureaus and expanded their internal IT organizations. The height of this centralized model was probably from 1979 to 1982—just prior to the introduction of the personal computer (PC) into corporations. With the introduction and spread of local area networks (LANS) and more PCs, IT groups started to outsource basic training, hardware maintenance, and some support. In the late 1980s the use of software packages increased as their capabilities rose and the underlying fourth-generation languages improved. Outsourcing of business activities began in this decade.

The 1990s saw the continued rise of the use of packages (a form of outsourcing) as well as expanded outsourcing of specific tasks and work. The use of contractor and consultant resources expanded. For some firms, more than 40% of the overall IT headcount consisted of contractors. The 1990s also experienced major attempts at outsourcing all the IT activities to one or several companies. Some of these failed miserably and the firms had to bring IT back into their organizations. In other cases, the firms began to contract out more narrowly defined activities. Companies also focused on core business activities and increasingly outsourced business functions.

What do these observations show? First, there is a substantial body of experience going back to before the late 1950s in outsourcing. This should be used to help you today. Second, several basic findings can be noted:

- Although outsourcing in IT may be worthwhile, outsourcing of other business activities may yield more benefits due to the size and nature of the work being performed. Mundane activities, such as shipping, telephone and utilities, and so on, are often outsourced. Many banks outsource their teller operations.
- It is valuable to develop a strategy and approach for outsourcing due to the complexity and interrelationships among IT activities. It is also important because of the requirement by vendors to have multiple-year agreements.

THE OUTSOURCER'S VIEW OF THE WORLD

What makes a firm enter outsourcing as a business? A firm decides to enter outsourcing in a particular area because they believe they have expertise in that area, possess excess capacity so that they can serve new or additional clients, and obviously make money doing it. The long-term nature of the contract is also an incentive, along with an almost virtual guarantee that unless they really screw it up or the business changes, the client will remain with them.

How does an outsourcer make money? They have to do the work that you do at lower cost, they have to provide added value, or achieve some combination of these. Here are some specific steps taken by outsourcers:

- Achieve economies of scale by sharing resources (hardware, network, facilities, software, people).
- Provide smarter or lower-cost resources to do the work.
- Charge for add-on requests and services that were previously done internally and taken for granted.
- Use more modern or more capable tools and methods that increase effectiveness and efficiency.

WHAT CAN BE OUTSOURCED?

Let's take a look at each area and discuss pros, cons, risks, and so on from the views of both the company and the vendor.

HARDWARE AND SOFTWARE RELATED

Operation of the Computer Center

The vendor will provide the operating staff and software to run the computer center if it remains an asset in the customer's books. Alternatively, the firm will sell the center equipment and other assets to the vendor. The vendor then pays the firm cash—very attractive if a firm is having cash flow trouble because the systems continue to function. Most vendors seek to merge the computer center equipment into their own data center. This supports economies of scale of staffing and networking. If the workload can be absorbed onto an existing application system, there are even more savings because some software licenses can be terminated with the release of the firm's hardware. Operating a computer center is a generic activity and does not involve company business processes—thus, making it attractive for outsourcing.

Some risks in outsourcing a computer center are as follows:

- What happens if the vendor goes out of business?
- Are your data and software secure?
- How will you oversee the computer center and ensure quality work?
- What if the vendor establishes proprietary software utilities and then runs your software using these utilities, making it more difficult and expensive to move off later?

- How will charges be established for additional services that require more hardware, software, or staffing?
- What are the liability conditions if the center fails?
- How will disaster recovery and business resumption be addressed?
- What are guarantees of service?

Outsourcing this activity is a long-term proposition because it tends to be very difficult to switch computer centers due to complexity. Not only must there be compatible hardware, but also there must a network conversion, facilities work, staffing up to take over, and the establishment of the same software environment.

Overseeing the vendor requires technical personnel (e.g., systems programmer and network manager/professional) as well as an IT manager. On-the-spot decisions that depend on funding approval are often required. You also seek to have service-level agreements in place where there are penalties if the firm does not live up to the agreed-upon service level.

PC Hardware Installation, Maintenance, and Upgrades

This may include filling orders for PCs, installing the hardware and software and connecting the PC to the network, troubleshooting if there is a problem, monitoring all software in terms of current licenses and serial numbers, supplying replacements when there is a failure, carrying out repairs, and upgrading the system or replacing it.

This is natural for outsourcing, except in areas where there is no support available. PC hardware is generic. The vendor staff probably works at a substantially lower rate than that of company employees. They typically are very experienced in doing the work—a factor that increases their efficiency. No internal company knowledge is required to do the work because it depends on the technology.

There are some risks involved in outsourcing. One is ensuring the quality of the work. A second is to make sure that the vendor continues to provide qualified staff. If a vendor has a locked-in contract for all your PC work, the vendor might taper off.

Here are some guidelines:

- Consider giving contracts to two firms for a 2- to 3-year period. Under this arrangement you reserve the right to call either one. Alternatively, you can set one up as a primary and another as a backup or alternate. You can split the work, with one firm doing installations and the other performing maintenance. They can back up each other.
- Set standards for response time after a problem is reported, and the mean time to repair a PC, and response on upgrades and other services.
- You will almost certainly need a help desk that can log and filter calls. Otherwise, you are likely to get embroiled in disputes over service calls.

Major Hardware Support

Almost all firms outsource the maintenance of their mainframe and minicomputer hardware. Most often, this goes to the manufacturer. There are standard maintenance contracts. However, beware that as your equipment ages, maintenance costs tend to rise significantly. This can be a major factor in deciding on upgrades and replacement.

NETWORK RELATED

Network Planning and Management

Network planning is important to the business. There is an ongoing role in measuring the network, determining what steps to take to improve performance, and planning for upgrades, replacement, and expansion. Unless an organization is quite small, this role is often performed by internal staff. This role also oversees contract work in installation and testing. The role has been expanded to extranets and linking to the internet as well as handling internal networks. Growth in internet use has made this position more significant.

Network Installation and Testing

You are not going to install and test a network every day. Network installation and upgrades are done as discrete tasks. Installation requires specialized equipment in terms of cabling tools, crimping tools, and so on, and network testing requires special monitoring equipment. These functions are also generic and not company specific, which make them suitable candidates for outsourcing.

Network Operation and Troubleshooting

With the increased reliability of networks, there is less of a need for internal staff to be assigned to troubleshoot network problems. For large organizations, it makes sense to have this function performed internally, allowing for identification of the problem prior to contacting a vendor to do repairs. The effect is to reduce overall cost and the mean time for repairs.

GENERAL SUPPORT

Help Desk

A help desk operation is where people respond to questions and problems from employees. You can have a help desk with a narrow scope responding only to systems and technology problems, or you can broaden it to include building problems, equipment problems, and any other nonemergency request.

Should you staff a help desk with employees? Clearly, these positions are junior level because almost all complex problems have to be referred to more technical people. It takes time to establish a help desk function. Start with defining the scope of the help desk. You need to find the people to whom you will refer calls. You will require time to build a database of frequently asked questions (FAQs), and software to track calls must be obtained. Then you must put the word out to employees that the help desk exists.

What are some of the risks of outsourcing the help desk? The basic concern is that quality of service is maintained. Another risk is that the estimated workload understates the actual workload by a substantial amount leading to increased cost.

Some suggestions for handling the help desk are as follows:

- Set up the help desk with internal employees first. This will provide you with a better understanding of the workload. In this way, you avoid the expensive setup costs of having a vendor establish the help desk.
- Implement a quality measurement program in which you follow up with people who have requested help to determine the level of service.
- Ensure that you have a personnel replacement condition in the contract so that if a contract person is not performing, there can be rapid replacement.

PC and General Computer Training

PC and general computer training include training in word processing, spreadsheets, graphics, electronic mail, and basic database management systems. In the early 1980s, companies conducted this training with internal staff. As that decade came to a close, most of it was outsourced to companies or local universities. Now, some companies require software expertise. Many others offer videotaped instructions, CD-ROM, or other training.

What mistakes do firms make in outsourcing training? One mistake is that they outsource training in more specific and technical software to general firms who just teach courses. This is part of a bigger mistake when a company wants to outsource everything to one vendor, often resulting in average to poor levels of training. For example, some instructors stand up and read materials. People could probably get more out of a manual. It is also very difficult to determine if the training is successful. People evaluate a class at the end, but then they may or may not use the material right away. Material not used for a long time is generally lost.

Application System and Other Training

There is a wide range of training that is possible. Examples lie in database management systems, the internet, general management training, and so on.

Another area is the software application system if you purchase a software package. For most of these topics, you would consider using outside help. Managers often make mistakes in outsourcing this training. Following are some of the pitfalls to avoid:

- The company specifies the training requirements only generally. The training that is delivered is not really suited to the company.
- The company fails to work with the training firm in the preparation and review of materials.
- There is no formal review of the training after it is given.
- The expectations of the company and the attendees are not matched by what is delivered by the vendor through a lack of forward planning.

MANAGEMENT SUPPORT

Systems Planning

Many firms are ill at ease or lack experience in developing a strategic technology plan. They may have failed when they last attempted it. Hence, there is often a desire to go outside for assistance. Managers think that it is useful to obtain an outside perspective.

There are some basic questions you should answer when considering outside help:

- What is the planning method of the vendor? Can you live with this method? Or does it seem too complex and complicated?
- How will the internal staff learn the planning method and gain from knowledge transfer from the vendor so that they can update the plan themselves?
- Should the vendor work in a joint team approach so that much of the legwork is performed by your staff?
- What are the planning end products? Are they specific actions that you can take, or are they general findings that require still further work?
- Can you place boundaries on the work performed so that the consultants do not run amok?

Technology Architecture and Assessment

You want to consider new technologies and develop an overall technology architecture. This requires external information and knowledge. It may include finding out what is going on with specific competitors and with the industry in general. It is clear that many organizations lack the internal expertise to undertake these tasks in a reasonable amount of time. Given that you would consider using a consultant here, some of the guidelines you would use to evaluate them are as follows:

- Find out if the consulting firms have any ties to specific technology providers. Do they have strategic alliances? This may bias their viewpoint.
- What knowledge and experience do they have of your industry specifically? What firms have they worked with in your industry?
- Obtain sample reports from their past work.
- Identify an outline and purpose of each end product so that there is agreement at the start about what is to be done. This will avoid problems later and head off the dreaded "scope creep."

SOFTWARE DEVELOPMENT AND MAINTENANCE

Turnkey Development and Customization

In turnkey development, a firm contracts for a complete system. It may be based on a package or be totally new. In customization, a vendor modifies an existing software package to add new features and capabilities. Turnkey systems were very popular in the 1980s with minicomputers. Their popularity has since declined with the expansion of standard software packages.

Systems Analysis and Design

These are areas where it is appropriate to bring in one or two people with specialized skills and knowledge about the new tools that will be employed in analysis and design. These people can serve as mentors to the internal staff. There are some potential pitfalls. First, your people are only exposed to one person's style and approach. Second, you could build up a dependency here if the people rely on the person too much. Another potential problem is how knowledge will be transferred to internal staff. Business departments may grow attached to these people and it may be hard to let them go.

Contract Programming

Many firms employ contract programmers. In some cases, the number of contract programmers amounts to 40% or more of the total number of programmers. Some contract programmers are former internal IT employees who retired or came back to address the needs of a specific system. Contract programming offers a number of advantages to firms. First, they can directly control what the person works on at a task level. Second, their costs may be less than internal programmers when overhead and benefits are counted. A third benefit is that management can change them out when necessary.

One disadvantage is that the firm gets dependent on the contract programmer and may not have a backup. The person may not be loyal or have a long-term interest in the systems work at the firm. They may want to learn some new technology. Either way, the firm is at risk. You should always assume that you must have some backup ability to recover if they leave.

What do the contract programmers do? In some cases, they perform maintenance—freeing up the internal staff for software development. In others, they have knowledge of new technologies and thus are employed in development.

System Integration

Given the demands on internal IT groups, it is not surprising that more firms turn to outside help for system integration. In some cases, system integration includes implementing turnkey solutions. More often, it means implementing large-scale network and software changes. If the current staff stopped every other activity, there still might not be sufficient resources. A system integrator brings a wide range of capabilities to the table. The major downside of outsourcing here is the cost. It is likely to be very expensive. Moreover, there will also be a high likelihood of dependence after the work is finished.

Data Conversion and Testing

In some cases, the data from the current system is migrated to the new system by writing temporary programs that move and reformat the data. These are one-time programs that are not needed at the end of the project. In other cases, there is a need for people to do data entry into the new system because the data in the current system is of poor quality or is incomplete. In both situations, there are outsourcing opportunities. The same applies to the situation when you are testing a complex system and lack internal staff to do the work.

Software Maintenance and Enhancement

Software maintenance and enhancement appear to be routine tasks. However, both require specialized knowledge of the application systems. If the system is old (say, for example, a 25-year-old legacy system), then it would be difficult to find people outside to do this work. What does this mean? Well, if you shackle the current programmers to the existing software, then who gets to implement the new systems with new tools? Not the internal staff because they are too busy maintaining and enhancing. It will be outside people who do the new implementation. This happens with many firms and raises substantial morale problems.

Some programmers see no future for themselves, except in maintenance. A few innovative firms attempt to use a mixed approach where outside people are used for both new development and maintenance.

SOFTWARE PACKAGE RELATED

Evaluation and Selection of Software Packages

Why go outside for help in evaluating and selecting software packages? There are several reasons. First, you would be getting an independent view. People inside the company may be biased toward an upgraded product that they already use. Second, a consultant has knowledge of and experience with different companies and industries with respect to the software package that cannot be found internally. Third, internal staff may be tied up so that it would take too long if they were to perform the evaluation. Also, they may not be as experienced in doing evaluations.

There are some dangers to beware of here. One is that the consultant may have experience with only one package. A more sinister danger is that the firm that the consultant works for has a partnership with a software vendor. In one natural resource firm, the accounting firm recommended two packages as finalists. Not surprisingly, these were the two firms that had strategic partnerships and alliances with the firm. Did the accounting firm tell the customer this up front? No. The customer had to find out on their own.

Implementation Support

For large, complex software packages such as *Enterprise Resource Planning* (ERP) software, it is almost essential to employ consultants and vendors to assist in implementation. The package is large and integrated. Previous implementation experience is a must. In such cases, the cost of the consultants can exceed the cost of the software by a factor of 9 to 10! The general advice here is that the consultants should be evaluated and selected at the same time as the software.

Due to the long elapsed time and high expenses, it is clear that the consultants must be managed very closely. A second challenge in management is to ensure that knowledge is transferred from the outside people to the internal staff. Otherwise, you risk long-term dependencies.

STEPS IN OUTSOURCING

Whatever you want to consider outsourcing, you should follow specific steps and a method to manage it and keep control. These steps are as follows:

- Step 1: Determine objectives. What do you really want to get out of outsourcing?
- Step 2: Identify and evaluate opportunities. Given the objectives, what are the best candidates for outsourcing?
- Step 3: Clean up the area or process to be outsourced. You cannot just go out and hire someone. You must measure what you have and clean it up to get it ready, like when you sell a house or car.
- Step 4: Evaluate and select the outsourcing firm.
- Step 5: Establish and shake down the outsourcing relationship.
- Step 6: End the outsourcing relationship and transfer the work inside.

Step 6 occurs naturally when you outsource development and some other activities. For activities where you have a multiyear contract, then you want to consider Step 6 as a contingency.

STEP 1: DETERMINE OBJECTIVES

What objectives do many firms cite when they consider outsourcing? Here are the top goals cited in several surveys:

- *Access to more talented and experienced IT people who know the technology.* The idea here is that if you can get these people on board, then you reduce the learning curve and speed up implementation.
- *Obtain a fixed level of services for a fixed price.* Related to this is the savings if the outsourcer buys your current hardware and then leases it back to you to give you short-term positive cash flow.
- *Competence of the outsourcing firm.* The internal staff and managers do not seem to be able to do the job — even if the cast of characters changes. Management gets so frustrated that they believe the only way that results can be obtained is to outsource it.
- *Management wants internal staff to focus on high-priority work.* They do not see any advantage in operating mundane systems activities that do not contribute to competitive advantage.
- *Save money.* This can be done by reducing the level of some services and through standardization with an outsourcing firm.
- *Provide new services that the internal IT group cannot.* This might be system development, new network implementation, system integration, and so on.
- *Obtain targeted help for a specific application or situation.* An example might be help in developing a plan.

Keep several of these goals in mind because you will be performing trade-offs in the next step and may elect one area over another based on the objective.

STEP 2: IDENTIFY AND EVALUATE OPPORTUNITIES

This step can be divided into specific actions. There are two parts to the evaluation: the evaluation in terms of the objectives of the first step, and a feasibility test to determine whether the activity can really be accessed from an operational and management view. At the conclusion of this step, you will have identified several areas that offer the most potential. Do not limit yourself to one because you still have not determined how the activities can be improved internally and which vendors are available.

Action 1: Identify Potential Outsourcing Opportunities

You might want to consider business processes that relate to IS activities. Examples are data capture, mail room operations, data entry, and customer information. The mistake that people make is that they decide on the specific activity when their overall goal is to save money. It is fine to be focused if you have a specific requirement, but you may miss out if you are attempting to reach a general goal such as those mentioned previously.

Action 2: Determine Evaluation Criteria

Evaluation criteria for outsourcing include the following:

- Potential improvement that the vendor could make in the activity in terms of performance or service
- Potential economies of scale through outsourcing
- Feasibility in doing the work without outsourcing; although negative, this criterion fits the situation where your outsourcing need is focused
- Interdependencies of the activity with other processes and systems
- Special knowledge required to do the work
- Barriers to outsourcing in terms of location, union contracts, and so on
- Dependence on specific technology that the vendor might not have
- Potential disruption in transitioning to the outsourcing vendor

Action 3: Evaluate Opportunities Based on Objectives

Here is what you can do to help evaluate your opportunities. Write down the objectives as column headings and the opportunities as rows. The table entry consists of how you rate the opportunity in terms of the objective on a numerical scale of 1 to 5 (1, low; 5, high). Ratings could be based on the following:

- Likelihood of achieving the objective with that opportunity
- Potential scale of benefit or savings
- Extent of time it will take to implement outsourcing

Note that there are factors to consider in terms of competitive position, management, and so on. These are coming next. Hence, the evaluation in terms of objectives is just a prescreening.

Action 4: Perform Evaluation Based on Operational and Managerial Feasibility

Just getting savings is not enough. When you evaluate outsourcing, you also must take into account the following:

- Politically, can the activity really be outsourced? It would, for example, be difficult for an airline that prides itself on its airline reservation system to outsource it. They would be the laughing stock of the industry.
- Organizationally, an activity might be tightly integrated into other ones. In such a case, the effort to split might be too great and too disruptive. An example occurs when the same group of people perform several activities. If one group performs development and maintenance, then you cannot outsource maintenance without dealing with development. Another example occurs when the activity is embedded in several critical business processes.
- The activity is involved in critical work that cannot be disrupted. An example was the situation in 1998 – 1999, when systems groups were heavily involved in fixing software for the Year 2000 problem.
- Ease and capability of being outsourced. If the activity is performed in many different locations, then outsourcing might be too difficult and time consuming. An example might be if you have four different locations of about the same size, each of which performs the same function. There might not be sufficient savings to address the activity in each of the four locations.
- The activity may depend on specific technology that should not or cannot easily be extended to an outsourcing vendor (e. g., access to a proprietary system and information).

Take the evaluation criteria of Action 2 and use these as columns in an evaluation table. Place the alternatives in the table in the rows. The entry in the table is the score or a comment about the alternative in terms of the criteria.

STEP 3: CLEAN UP THE AREA OR PROCESS TO BE OUTSOURCED

Do not assume that a business or an IT activity is efficient. In the case of Roberts Agency, the customer information activity was viewed as the best candidate for outsourcing. The customer information area depended on an ancient system, along with manual procedures. The telephone system was antiquated. Staff turnover was high. There had also been a parade of managers. It sounds perfect for outsourcing as is, right? Wrong. An outsourcing vendor could charge a fortune to do minor improvements and then lock you into a long-term contract. The approach that was taken was to give the area some attention to improve it and measure it. Then it was put up for outsourcing.

What was done to clean up customer service? An off-the-shelf logging and tracking system was acquired without buying additional hardware. Standardized procedures were put into place. Staff training in customer service was conducted. Process measurement was also implemented. There was a substantial improvement. Costs were reduced by 15% and turnover was cut by 25%. Call volume handled rose by 20% for the same staff. Why outsource it then? The costs were still too high. The activity was not viewed as a core activity. Personnel costs were higher than private industry.

What are some constraints that you face when you fix up an activity? We answer this before you address how to clean it up:

- There is only a limited time for clean up. Usually, this is 3 to 6 months.
- You do not have time or money to embark on any major automation project. Anything you do will likely be replaced when the activity is outsourced.
- You have to be very careful in making any promises to people involved in the activity because they might be part of the outsourcing. Consult with the human resources department on this.

With these constraints what can you do? From experience, here are some potential areas of action:

- *Policies.* You may be able to simplify or streamline policies affecting the activity. If the activity is to be outsourced, you will certainly want to install new controls over the incoming workload.
- *Procedures and workflow.* This is the nuts and bolts of who does what and how they do it. Making procedures easier will likely reduce costs later.
- *Training.* You can update the training of the staff in the activity.
- *Documentation.* The procedures and methods must be documented and turned over to the outsourcing vendor.
- *Minor systems and technology work.* Although you cannot do wholesale replacement, you can consider enhancements in the software tools.
- *Measurement.* You definitely want to install measurements of the activity.

What have you accomplished with this effort? First, you have lowered costs and increased efficiency in the short term. It is possible that the results are so dramatic that there is no need to consider outsourcing. A second impact is that the activity is now better understood, documented, and measured. This will become very useful when you are negotiating with an outsourcing vendor. The vendor cannot claim that the first thing they have to do is clean it up. Overall, by making it more efficient, you are in a better position to perform the following:

- Negotiate on price and other terms
- Measure and control the vendor effort
- Support a faster turnover to the outsourcing firm

During this step you will have to work with human resources on what information will be provided to the staff performing the activity.

In some cases, the clean up work is very specific. For example, if you want to outsource maintenance and enhancement, then you might undertake the following:

- Review and upgrade the documentation of the application systems.
- Review all outstanding requests and eliminate any that lack clear tangible benefits.
- Make some minor programming changes to fix current problems.
- Prepare a walkthrough presentation of the system for vendors.

If you are considering outsourcing computer operations, then you might do the following:

- Review all operational systems as to existing production problems.
- Attempt to fix some or all the production problems.
- Review the operations, security, and other procedures and update as needed.
- Determine if there are operational procedures that are followed informally and not documented, and then document, these.

STEP 4: EVALUATE AND SELECT OUTSOURCING FIRM

Action 1: Identify Potential Vendors

You can find vendors through IT managers, purchasing, the web, and literature. Make an initial list and have these reviewed by managers. Then draw up a list of questions that can be posed to the vendors to narrow the field. Some questions are as follows:

- What expertise does the vendor have relative to the specific activity?
- What resources are available to support the activity?

Gather information from the vendor.

Action 2: Develop a Request for Proposal for Distribution to Vendors

The request for proposal should contain the following:

- Goals of the outsourcing
- Description of the requirements for outsourcing work in the form of lists
- Skills and knowledge required of the outsourcing vendor
- Technology and systems employed in the activity
- Measurements of the current activity
- Transition approach that will be employed after a vendor is selected and contract signed
- Ongoing measurement and management control approach
- Evaluation criteria that will be used for proposals
- Evaluation method that will be employed for the proposals.
- References of current customers of the vendor

Action 3: Issue the Request for Proposal and Conduct Vendor Conferences

After issuing the request for proposal, you normally hold a bidders' conference. This is an opportunity for you to present the objectives and additional information related to outsourcing. It is also a chance for vendors to ask questions. What do vendors typically ask? Here are some examples:

- What is the term of the outsourcing engagement?
- Of the objectives stated, how do these rank? In particular, is the major objective to save money?
- What do you envision happening to the internal staff?
- What restrictions and methods will be used to evaluate the current staff in the activity?
- What documentation exists related to the activity?

Action 4: Evaluate Proposals and Select Vendor

You can employ the criteria defined earlier as a starting point in the evaluation. Early on, you should contact customers of the vendors that are identified in the proposals. Assuming that they are doing outsourcing in the same area as you are considering, here are some questions to ask:

- How long have they had the outsourcing agreement in effect?
- What was the timeline of events from initial evaluation to today?
- What were their original goals for outsourcing?

- Did they achieve these goals? If so, how? If not, why not?
- How do they manage the outsourcing vendor?
- How do they handle additional work requests?
- What is the quality of people provided?
- Has their staffing and management been stable? What is the turnover?
- If they had it to do all over again, what would they do?
- Would they employ the same vendor for additional work?
- What benefits did they expect and achieve with respect to outsourcing?
- What were the difficulties and challenges that they encountered?

After you receive the proposals, the normal method is to evaluate them and select a handful for further consideration as finalists. The narrowing of the field is often based on elimination due to lack of experience.

Action 5: Negotiate the Contract with the Vendor

Space does not permit a detailed discussion of contract negotiations. In addition, negotiations are discussed in a later chapter in regard to software packages. Here some critical success factors will be highlighted:

- *Dispute resolution process.* How will problems be handled?
- *Management process.* How will the outsourcing relationship be managed and coordinated?
- *Measurement.* How will vendor performance be measured?
- *How additional unplanned tasks will be handled.* New opportunities will almost certainly arise. When this occurs, what is the method for having these opportunities addressed?

During this action, the firm must identify the project leader who will establish the outsourcing relationship and monitor the vendor.

STEP 5: ESTABLISH AND SHAKE DOWN THE OUTSOURCING RELATIONSHIP

Once the contract is established, you and the vendor must work out a transition of the activity. This involves human resources, several departments, the staff, and the vendor. Given the number of players and potential for problems and misunderstandings, it is recommended that you develop a project plan. Begin with identifying the milestones associated with the transition. Make a list of issues with the players and associate them with the milestones. Examples of milestones are as follows:

- Determination of how staffing of the activity will be done; will the current internal staff be transitioned to the vendor?
- Definition of the technology to be employed in the new situation and the interfaces required
- Definition of how the activity performed by the vendor will be measured and evaluated
- Determination of the approach for resolving and escalating issues and disagreements
- Simulation of the transition with the vendor and human resources
- Development of guidelines and procedures for staff through human resources
- Actual migration of the activity
- Initial monitoring after transition

Through building the project plan and identifying and dealing with issues, the firm and the vendor can build a working relationship.

STEP 6: END THE OUTSOURCING AND POTENTIALLY MOVE THE ACTIVITY INSIDE

As indicated earlier, this step may be a natural one for certain activities. For others, it is a contingency that you have to consider. Whichever it is, you should work with the vendor to develop a transition plan to move the activity back inside. Such a transition plan might include answering the following questions:

- What documentation will be needed by internal staff?
- Can outsourcing staff be considered for employment?
- How will internal staff learn about the activity?
- How will knowledge gained in outsourcing be transitioned to internal staff?
- How will the transition be managed?
- How will the activity be phased down for preparation for transfer?

WHAT CAN GO WRONG?

- You fail to clean up the activity prior to outsourcing. This can occur if you are under severe time pressure. If this happens, then you should work with the vendor to define a clean-up approach. This approach should define specific goals as well as methods for the work.
- The vendor fails to perform adequately. There could be a misunderstanding in that you and the vendor have different views on their performance. They

may blame internal staff for not making information available. They may even state that internal employees are trying to sabotage the effort. To head off a crisis, keep track of issues. Make each meeting with the vendor manager focus on solving problems and issues. Develop a measurement method for what they do and measure the activity on a monthly basis to determine performance. For some activities such as training, this may mean conducting surveys of people with whom the vendor works.

- The vendor works with your staff and business staff to try and expand the outsourcing. This is not unexpected. The vendor does want to have the outsourcing make money and so naturally would like more work. Do not say no to this, out of hand. Instead, set ground rules at the start for how new work will be evaluated. Indicate to the vendor the limits on contacts and marketing efforts.
- The vendor subtlely resists making the transition back to internal staff. The vendor wants to keep the work. They may be reluctant to cooperate. A good approach here is to apply pressure through daily meetings, if necessary, to develop the transition plan, supply documentation, make available staff, and so on.

REDUCED SCHEDULE AND COST APPROACH

How do you try to reduce the amount of time and effort to get outsourcing going? Here are some suggestions:

- Carry out identification of vendors in parallel with the activity clean up.
- Do clean up until the time that the transition starts. This will lower the cost and potential for problems later.
- Prepare documents and line up people for the transition in parallel with negotiations.
- Work with human resources very early in the process to get them involved. If they do not give it a high priority, then you could experience delays.
- When the vendor has been selected, get the transition approach and planning started immediately. This tends to be a source of delay.
- Jointly develop a transition project plan with the vendor.

EXAMPLES

Vision Insurance outsourced some of their customer service and claims along with some of the staff that performed these functions. They were in such a rush to outsource and "save money" that they did not attempt to make the areas more efficient. They did not do Step 3 in our sequence. As such, they really had no idea of what the savings should be or even what customer service should cost to operate. They selected a vendor without regard to what systems would be employed. Their decision criterion was low overall cost. It was a disaster. The vendor used a proprietary system that required special systems interfaces with the internal agency systems. Do you think this was included in the price? Forget it. It was another added cost item.

The closest that Vision Insurance could come to estimating benefits pointed to savings of less than 2%. Moreover, customer complaints about services grew. Finally, Vision Insurance decided to bring the functions back into the company. There was a substantial cancellation penalty in terms of staff, software conversion, and facilities costs. This example demonstrates the need for defining objectives and cleaning up the activity prior to outsourcing.

A parts distribution firm outsourced almost all IT activities to three firms. There were objectives and IT activities were streamlined ahead of time. Internal staff were transitioned to the outsourcing firm. This appeared to be a real success. In fact, magazine articles appeared touting the approach. Unfortunately, they left out the last step. The company did not allocate enough resources to manage the outsourcing firms and their interrelationships. Misunderstandings occurred, costs escalated, and problems and issues remained unresolved. Finally, one of the contracts was canceled and the other two contracts were narrowed in focus.

A Brazilian financial group outsourced a number of its IT activities. The following were lessons learned based on their experience:

- Cost savings that were at first viewed as primary were later viewed as being of less priority. Major benefits were the access to additional resources, new ways of thinking about projects, and improved control.
- Knowledge transfer from the vendor must be managed and directed; otherwise, it will not happen.
- Standard contracts from vendors should not be used except as a starting point. Contracts must clearly state objectives and measurement standards.

LESSONS LEARNED

- Where possible, identify more than one vendor for the work. Consider dividing the work among several vendors so that you reduce your dependency.
- Review all existing outsourcing agreements to assess how effectively they are being measured and managed. Consider developing an overall ranking of vendors.

- Encourage internal staff to share experiences and lessons learned from outsourcing. This will raise their skill levels and make them more effective in dealing with vendors.
- Seriously revisit existing outsourcing arrangement and evaluate what it would take to terminate the relationships. This will give you knowledge about the status of the outsourcing.

SUMMARY

Outsourcing is now a more frequent action by IT groups. However, surveys indicate that almost half of all companies are either partially or wholly dissatisfied with the process and/or the results. Part of the reason for this stems from the lack of a structured approach to outsourcing. In particular, there is insufficient attention given to the clean up and preparation step presented earlier.

A second problem area is the lack of a formal approach for managing the relationship with the outsourcing vendor. The internal project manager who is assigned to oversee the vendor often has other duties and management does not place sufficient value on the outsourcing.

WHAT TO DO NEXT

- Review an activity that is currently being outsourced. Develop a score card for the activity that measures the vendor performance. Evaluate the extent of time spent in overseeing the outsourcing and quality of management.
- Companies often fail to capture lessons learned from their outsourcing experiences. This then leads to the same mistakes being made repeatedly. Set out and capture lessons learned related to each step of the outsourcing as defined in this chapter.

Part II

Software Development

Chapter 5

Define Requirements and Do Prototyping

INTRODUCTION

In Part II, you develop application software based on the requirements identified for the new process in Chapter 3. Chapters 5 through 7 cover the development of the software and integration, whereas Chapters 8 and 9 encompass training, conversion, procedures, and operations support. The scope of this chapter includes defining detailed requirements, selecting methods and tools, and developing a prototype system. This prototype is then tested with the new process in the pilot to validate what has been done. Chapter 6 follows up on this in terms of full development and integration from the prototype.

Prototyping has been used by industries for years. By building a prototype, you can evaluate the features and characteristics of the model. In information technology (IT), prototyping was almost impossible in the late 1970s. There just were no tools. In the 1980s, "phony prototypes" that simulated a user interface could be built. However, there were a number of problems with these. First, they did not test any transactions. Second, when you were done, you threw the prototype away and started over again building a system. Even so, the benefits of prototyping were so substantial that this prototyping continued.

The software tools have now advanced to allow you to construct prototypes that work and that later can become part of the production system. The benefits of prototyping include the following:

- General graphical user interface (GUI)
- Validation of job functions through menus associated with each job
- Verification of the data elements for screens
- Screen layout and design
- Flow of transactions by using status codes and workflow queueing
- Evaluation of the new process through the prototype
- Estimation of benefits potential for the new process

There are seven steps to constructing a prototype. They are identified as follows:

- Step 1: Identify the development team and establish the development plan.
- Step 2: Develop detailed requirements for development.
- Step 3: Determine the methods and tools to be applied throughout development.
- Step 4: Perform data analysis.
- Step 5: Develop the first prototype.
- Step 6: Conduct a pilot of the prototype and the new process.
- Step 7: Refine the prototype and conduct more pilots.

There is some overlap in these steps. That is, you cannot hope to complete Step 2 before you work on later steps. This occurs because there is too much detailed work to do, and thus the business rules will not be completed until the latter half of programming. After you begin Step 2, you should proceed with work on Steps 3 through 5.

Because this is the real world, you must constantly trade off where to spend energy, time, and money. In an academic setting, you could spend a great deal of effort in documenting requirements and design without ever programming. Not so in real life. There is constant, unceasing pressure to show results. If you spend too much time on design, then the business department may lose interest. If you do not do any design work, you will technically fail. Hence, the trade-offs will be identified as work progresses, with the goal being to do as little supportive work as possible. First concentrate on the mainstream of the prototype and then the pilot.

Another guideline is that the system you build should be as simple as possible. Do not aim for elegance. Try to reuse code as much as possible. Steal code from other programming efforts. Do the same with design. The goal is to complete development as soon as possible within the budget—not to build some exotic solution. This has implications for your team in both design and development:

- If a document is going to be read by department staff, it must be understandable and not composed of exotic charts.
- Target producing as little documentation as possible. Anything you document consumes time to write, review, and update. Take a zero-based documentation approach.
- Try to get business staff to do as much as possible in terms of documentation, gathering business rules, and so on.
- Break work into small chunks—sections of a manual, individual parts of a design, and individual programs for management control.
- Regardless of whether or not you use object-oriented tools, you should use an object-oriented approach. This will make the work more modular and measurable.

STEP 1: IDENTIFY THE DEVELOPMENT TEAM AND ESTABLISH THE DEVELOPMENT PLAN

THE DEVELOPMENT TEAM

We begin with a few lessons learned from past development projects:

1. Keep the core development team as small as possible. The larger the team, the more cumbersome and time-consuming project coordination will be.
2. Get people on the team as late as possible. You do not know what you need when you start. You do not want people sitting around. Therefore, only recruit people when you have defined in detail what skills they need to have and what tasks they are to do.
3. Involve as many department staff as possible in the development on a part-time basis.
4. Once you develop the plan, stick to it. Changing the plan based on rather small issues will create morale problems.
5. Be steadfast in adhering to the methods and tools that you select. Changing these in midstream can cause disruption to the project.

Consider the second point. Traditional wisdom is that you should get critical business users on the team who have a wealth of knowledge and dedicate them to the team full time. That is, they will be the source of business knowledge. This point is wrong and illogical. First, given complex business processes, no two or three people know it all. Second, when you take away critical people from the department, the department suffers. Furthermore, you incur the hostility of the department management. Third, by involving more people you get different points of view. Fourth, by involving more people, you increase the audience of support for the project.

Who do you need to start? You may need a systems analyst besides yourself to gather requirements. You will also require a coordinator from the business department who can help gather requirements. Later, when you are about to develop the prototype, you can add the programmers. On the business side, you can increase the number of staff involved in business rules as you go. You can even involve senior people if you have targeted tasks for them to do.

THE PROJECT PLAN

Some specific guidelines for a system development plan are as follows:

- Be able to relate business rule definition to specific tasks in the plan. This will make the plan much more detailed but will give you traceability.

- Make sure that you have a regular stream of milestones. Avoid having any large time gaps without deliverable items.
- Each computer program, no matter how small, should be included in the project plan at the detailed level. This will support configuration management and provide you and others with a better picture of the project's status.
- Each person on the team should define and update their own work. This helps ensure accountability.

Additional guidelines on project management are provided in Chapter 9.

STEP 2: DEVELOP DETAILED REQUIREMENTS FOR DEVELOPMENT

In Chapter 3, you developed a series of tables as follows:

- For selected major transactions, the columns were transaction step, old process, new process, and requirement.
- For transactions defined, the requirements and any comments comprised a second table.
- A third table identified the requirements, why they were needed, and comments.

Now you must get down to the detail. Because you and the team are defining a new process, the level of detail is more than can be done sequentially in this step. In other words, the work in this step will continue throughout later steps.

ACTION 1: MAKE A COMPREHENSIVE LIST OF TRANSACTIONS OR TYPES OF WORK

Let's start with the end product. You want to end up with the table in Figure 5.1. The description indicates the purpose of the transaction. The type of transaction refers to entry, update, reporting, and so on. Frequency is the number of transactions that occur per day or in some other period. The volume of the

Transaction	Description	Type	Frequency	Volume	Performance	Interfaces	Comments

Figure 5.1   Transaction table.

transaction refers to the number of the transactions that occur over a longer period. Performance pertains to the expected response time needed to handle the transaction. Interfaces allow you to indicate what external systems the transaction requires. Additional information can be added as comments.

What purpose does this table serve? Here are some applications:

- You can organize the transactions by type. Each type can be considered to be a class of transactions. This will help in development, where the code for a transaction can be reused for other transactions in the class.
- Batch processing can be placed in a similar table.
- Based on frequency and volume, you can prioritize the interfaces that will need to be developed.
- The table can be employed to set priorities as to which transactions to include both in the prototype and for development.
- Performance can assist you in determining programming objectives.

ACTION 2: FOR EACH TRANSACTION, IDENTIFY THE BUSINESS RULES ASSOCIATED WITH IT, ALONG WITH THE STEP DETAIL, INPUT, OUTPUT, AND INTERFACES

Figure 5.2 provides a table for a transaction. For each step, enter the description, input, business rules, output, interfaces, detailed requirements, and comments. This table helps in both the prototype and production development. As in Action 1, batch transactions and work can be defined in a similar table.

Transaction: _____

Step	Description	Input	Business rules	Output	Interfaces	Detail requirements	Comments

Figure 5.2 Transaction detail.

ACTION 3: DEFINE THE EXACT ROLES OF THE STAFF IN THE BUSINESS DEPARTMENT AND WHAT THEY CAN PERFORM IN THE BUSINESS PROCESS

At the core of this action, you are defining the future job duties and descriptions of the staff. Who can perform each transaction? Who performs the

Transaction: ─────────────

Step	Job type	Comments

Figure 5.3   Transactions and job types.

individual steps in the transaction? Why are answers to these questions important? A business process is defined by the details of the work—transactions and steps. People's job functions and descriptions for the process are built from this detail. If you just write down what people do now, you will probably eliminate the possibility of achieving major benefits because the organization of the work among the people will determine both effectiveness and efficiency.

Figure 5.3 provides a table in which to enter this information. The columns of the table for each transaction are step, job type, and comments. You will be inventing new job titles for the job type. Skills refers to the knowledge, experience, and expertise that are required. This will help in associating current staff and positions with the new ones. Additional information can be placed in comments.

How is this table used? First, you can employ the table to help define menus for each job type. Second, you can begin to determine if the benefits that were originally estimated might come true. This is significant and can be shown to upper management and to the department manager. If shown to department staff, indicate that this will show them how their work will change.

ACTION 4: GATHER ALL FORMS, REPORTS, AND OTHER DOCUMENTS AND MAKE A LIST OF DATA ELEMENTS OF THE CURRENT SYSTEM AND ADD ONES FOR THE NEW PROCESS

Gather up all forms. These were identified in the input part of the transaction table in Chapter 3. You might use the table in Figure 5.4. In this table, you would list the data elements for each form, how the data element is created, how it is used, and comments. You would do the same for reports. Also, gather data from the current system to identify the data elements. Finally, try to find shadow

Form: ─────────────   Identifier: ─────────────

Data element	How created	How used	Comments

Figure 5.4   Data elements for each form.

Table	Purpose	Transactions	Comments

Figure 5.5   Tables for the new business process.

systems and workarounds and see what data elements are employed. Enter these data into a similar table.

Turn to the tables that support the processing of the work. These tables might include sales tax, zip code–city lookup, and so on. Use Figure 5.5 to record this information.

Look around for any logs that are maintained by the staff to record the arrival and completion of work. You can use Figure 5.6 for this. What does a log reveal? It tells you how people track the work manually. You will want to eliminate these logs for productivity. This will also help in determining workflow tracking and status codes.

These tables will serve as the basis for designing the databases and files. You can add data elements that business staff want but that are not in the current process. Review these tables with business staff. Just looking at lists of data elements or even diagrams can be extremely boring and may not elicit feedback. Use the forms, logs, reports, and so on, in the meetings as props (such as in a movie) and walk through transactions.

Log	Purpose	Who maintains	Transaction to which log applies	How log is used	Comments

Figure 5.6   Table for identifying logs.

ACTION 5: DEFINE INTERFACE REQUIREMENTS

In the earlier tables, you identified which systems the transactions in the new business process will interface with. Use the data to construct another table (Figure 5.7). This table provides the transactions, nature of the interface, volume, batch/online, and comments. You can summarize this in Figure 5.8.

System: _____

Transaction	Nature of interface	Volume	Batch/online	Comments

Figure 5.7   Interface requirement for a specific system.

System	Summary of interface	Characteristics	Comments

Figure 5.8   Summary interface requirements.

ACTION 6: IDENTIFY HOW WORK WILL BE TRACKED AND MEASURED IN THE NEW PROCESS WITH THE NEW SYSTEM

Using the logs and direct observation of the process, you seek to identify how work is tracked today and how the new process should be tracked. You can do this based on transaction. Include the status code, the step of the status code, the action or trigger required to change the status code, where the transaction goes next, and comments. The table is shown in Figure 5.9.

Transaction: _____

Status code	Step in process	Action/trigger	Where next	Comments

Figure 5.9   Status codes.

ACTION 7: DETERMINE THE TECHNOLOGY THAT WILL BE REQUIRED IN DETAIL TO SUPPORT THE WORK

You made a stab at determing the technology in Chapter 3. Maybe it is still okay. However, you now have much more information and can identify requirements for such things as the following:

- Mobile communications
- Scanning
- Electronic data interchange (EDI)
- Electronic commerce

Now you want to identify security and access control requirements, as well as production and network requirements. This action is important because it should be the last time that you consider the technology for the new system.

To help you to evaluate the technologies, make a drawing of the hardware and network components. Do the same for software. Label all information flows and connections. This is your system architecture and assists you in determining if you are missing pieces.

Carrying out these actions has additional major benefits that go beyond the information and support for development. These benefits are as follows:

- The tables assist in validating the benefits of the new process.
- The tables can be used to start to define procedures for using the new business process.
- The tables can be employed to start collecting test data.

Perhaps, most important, they show the business staff what the new process will really be like.

STEP 3: DETERMINE THE METHODS AND TOOLS TO BE APPLIED THROUGHOUT DEVELOPMENT

Often, people approach determing the methods and tools in a backward manner. They start with the tools and let them drive development and the methods applied. Do not fall into this trap. Begin with defining how you will do the work and then select the tool. These are the critical areas where you must make decisions. Space does not permit going into individual methods or tools. There have been literally hundreds of these invented since the late 1960s. Some basic selection criteria lie in answering the following questions:

- What does the IT staff use now?
- What are the benefits of the method to development?
- Will the results of using the approach be maintainable?
- How much effort is necessary to learn and use the method?
- How will the method be enforced?
- How will results be reviewed?
- How will you convince the IT staff to use the new method or tool?

These are not idle questions. In fact, many firms have adopted methods and tools without answering them. Sometimes, in less than a year, the method or tool is quietly dropped. That is why experience gives you the following guidelines:

- Use as few methods and tools as possible. The more you use, the greater the effort spent for the methods and tools and the less spent for development.
- Concentrate on tools and methods that are self-motivating and self-enforcing.
- Contrary to vendors, there is no silver bullet method or tool.

Experience shows that the most productive areas for tools are those involving programming, database design, integration, and testing. Object-oriented methods have also been found to be successful.

Intranet development using Java, Active Server Pages, or similar tools has been said to also be successful in that it tends to be self-enforcing because the structure of the programs and size is controlled by the tool itself and the internet environment. An intranet/extranet application, although not appropriate for every online application, has several advantages. These are as follows:

- There is less user training because many of the business staff are already familiar with web browsers. Training materials are also reduced.
- In production, with the proper security, business staff and managers can access the system remotely.
- There are more tools and aids available for intranet development today. More are coming on the market.
- The programming is modular so that it is simpler to make changes.
- Because of the modularity, it is easier to test.
- The hardware and system software resources required for intranet development are limited.
- With an intranet development project, it is possible to do development in teams over the internet.
- With web hot links, the design and programming are easier and more compartmentalized. There is less need for hierarchical menus.

Following are some myths associated with methods and tools:

- **Myth 1: Documentation tools are very useful in helping the business staff to understand requirements and design.**
  This is a nice idea until you try to apply them. The symbols are not intuitive and the business staff has difficulty in understanding the output. Creating the documents also takes a great deal of effort—time taken away from analysis and design.

- **Myth 2: You can automate the requirements design and code generation.**
  Many tools have tried to do this. Few survive today. It is true that you can use programming libraries to aid development. Automated tools, however, often do not include exceptions or provide an overall understanding.

- **Myth 3: The tools improve programmer productivity.**
  This is not necessarily true. Programmers have their own tools. If they write programs using modern object-oriented languages or fourth-generation languages, then these have their own rules and methods. They would have to take the output of design and analysis tools into useful information. If the work is carried out at the transaction level, then this is too detailed for many of the methods and tools.

STEP 4: PERFORM DATA ANALYSIS

To construct the prototype, you must perform work to define the databases as well as supporting control and parameter tables. Here it is useful to develop the database design and normalize the databases. After defining these, you must validate them by seeing how transactions interact with the databases and tables. In terms of tools, data dictionaries and the database management system itself are useful and generate documentation. Review these with the IT staff.

STEP 5: DEVELOP THE FIRST PROTOTYPE

Having selected the methods and tools, you are now ready to start building the first prototype. What do you hope to accomplish with this?

- The business staff will see the GUI for the first time.
- They will see the menu structure for each job type. This will be important because it will translate into their future roles and responsibilities with respect to the business process.
- The implementation project will receive a big boost in credibility.
- This is an effective way to motivate the business staff to continue defining the business rules and transactions.
- They will actually be able to see one transaction appear to be processed in terms of local databases.
- The feedback from the business staff should be encouraging to the developers.
- The benefits of the new process will receive further validation.

This is a tall but reasonable order. What will you actually do in the prototype? Here is what you should include:

- Logon screen and security for user IDs and passwords
- Processing of the user ID to display the menu appropriate to that job type; the job type is linked to the user ID
- The menu for each job type
- The ability to enter one transaction on a data entry screen
- The ability of a supervisor to review the transaction
- Support for the workflow and status codes associated with the transaction
- The linkage of the transaction screen to a database; this allows for processing of the transactions without the business rules and supporting workflow routing of transactions

In traditional online, character-based systems, this would be a tall order. Even with clientserver, this can be a substantial effort. However, with intranet technology, you can use the browser interface and then apply standard buttons, toolbars, and pull-down menus. Development has the advantages listed in a previous step. There will also be less set up of software to show the system to staff. In client server, you would have to load the client software on each personal computer (PC). Because many PCs have a browser, the work is greatly reduced.

Why not do more? Because you want to find out if you are on the right track. You do not want to have to undo it all later. Also, you do not want to spend too much without showing the business staff some progress. A prototype requires planning for success in a limited amount of time. Here are some guidelines:

- Select the initial transaction based on a common form, if possible. The staff are familiar with the form and with handling it. They will see the benefits of the system faster.
- Show the prototype to a few selected business staff first. Get their reactions by asking the following questions: Is it easy to use? What would they like to see changed? What additions would they like?
- After making some improvements to this prototype in terms of tuning, you can show it to many more business staff. Some hints are as follows:
  - Have the business staff who tested the first version demonstrate this version. This is more credible than the project leader or systems analyst doing the demo.
  - Introduce the prototype by stating the limitations and indicating that the road to development will be long and hard.
- Do not produce any documentation; the prototype tends to document itself. Also, any time spent documenting takes you away from other activities.

How long should it take for the prototype to be developed? It varies based on the tools and skills of the people. A reasonable estimate is 4 to 6 weeks.

How do you manage the development of the initial prototype? You manage it very closely. You want to employ this step to get the team into a work pattern of sharing information. You also want to establish standards for the GUI, as well as the initial programs that can be reused as you do later prototyping and production development.

STEP 6: CONDUCT A PILOT OF THE PROTOTYPE AND THE NEW PROCESS

After some refinement with the first and later prototypes, you can demonstrate the system in the context of the new business process. That is, instead of showing the system by itself, you will show the new business process. The system will just be part of the process. This demonstration and evaluation is called the *pilot*. Some guidelines on doing a pilot are listed:

- As with the first prototype demonstration, the business process should be shown to the business staff by business staff involved in development.
- Create a half-page form as an evaluation sheet. The sheet should contain job title today, number of years with the current process, what they liked best about the new process, and what they want to see improved.
- Conduct the pilot several times in a conference room. A good time is at lunch. Serve cookies and soft drinks.
- To demonstrate the process, you must show the manual steps and any paper forms or logs used. The demonstration is similar to infomercials on television. It must be complete to be credible in terms of the process. If you gloss over some parts of a transaction, people will catch on and the credibility of the prototype will suffer.
- After the staff has demonstrated the process, select members of the audience and have them do a transaction.
- Make no changes between demonstrations. Keep it stable.
- Try to avoid having managers and supervisors in the demonstrations of the process. Their presence can be intimidating. Instead, give them a preview demonstration and explain what will happen. This will hopefully dissuade them from attending the later demonstrations.
- Do not exhaust the audience in the first pilot. There will be more pilots. Point out that each person should see two versions of the pilot process so that they can see progress.
- With each successive pilot, explain how you incorporated their comments and what new features have been added.

What are the benefits, and what do you hope to achieve with a pilot?

- You will establish credibility for the new process. As a result, the department staff will be less intimidated and fearful of the new process.
- People will realize that the system is just part of the process.
- They will see the importance of gathering transactions and business rules.
- Enthusiasm should grow.
- Obviously, you will get feedback to help with successive prototypes.

What are some of the risks, and how do you mitigate these?

- Expectations may be raised when "department staff" see something tangible. Make sure you dampen these expectations by indicating that there are many more tasks to do. Focus on the tasks that the business staff will have to do as well as mentioning the programming.
- The audience may suggest changes that are not practical or that have been previously considered. Give people credit for the ideas and keep going. You want to avoid the situation where there is never a system—just continuous prototyping. People can become tired if there is no system. They will likely not want to be involved.

- The programmers may want to keep improving the prototype in nonessential ways. This is a natural impulse because they are getting positive feedback from the business staff. Suppress this. Aim at implementing only the essentials.
- The business managers and staff react positively and now suggest many new requirements. Put these into change control and make them issues. You will drop any that do not pertain to the new business process. You will also eliminate any that are nice to have but that yield few benefits and that they can live without.

STEP 7: REFINE THE PROTOTYPE AND CONDUCT MORE PILOTS

Successive prototypes should include the following:

- Add to the number of transactions without business rules.
- Provide more workflow logic for status codes and routing of transactions.
- Produce a few reports for tracking of work.
- Create some canned operational reports.
- After the above steps, you can add some business logic along with control tables.
- Add interfaces to spreadsheets, word processors, electronic mail, and groupware.
- As the prototyping continues, begin to freeze elements of the system. First, you should freeze the general GUI. Then you will fix the menus and navigation of the system. Later, you will freeze the data elements; otherwise, the prototype can get out of control.
- The workflow routing of the transactions will most likely grow more complex because there can be incomplete or rejected transactions that have to be reworked.

As you are prototyping, you will be collecting changes and extensions to the prototype. Spend time organizing and analyzing these. Conduct a working session with the programmers to determine the order of implementation. When you are about to embark on a new transaction, have the programmer explain how the current code can be reused. Hold meetings with programmers at least several times per week. It is important to build momentum. At some meetings, conduct a walkthrough of the source code. This is especially important for the first few transactions because the code will serve as objects from which successive transactions will be derived.

Exercise configuration management control of the software. That is, ensure that there is a development library and that programmers know what each program is working on. At the end of the prototype, you should have the client part of the system defined and nailed down. Hopefully, you have selected methods and tools that do not require you to dump everything and redo the work in the production development setting. Note that several things have not been included in the prototype. These include extensive business rules; links to external; host systems; batch programs; and production backup, recovery, and restart programs.

From experience, you are likely to make the following types of changes:

- The original design and specification of job types will have to be changed. As more people see the pilot, they will give input on the details of how the process works.
- The data elements that you believed were complete are probably not and you will have to add some.
- New transactions will surface. These are the ones that are not supported by the current system but that the department does as part of the process.
- Manual parts of the process will surface and cause some expansion of scope. However, if these are essential to the process, they should be included.

WHAT CAN GO WRONG?

- People can be consumed by the urge to get the requirements right before doing development of the prototype. This can delay development. Instead, keep in mind that the prototype and the pilot will test the requirements and their validity.
- The prototype must be cut off; otherwise, there is no stability. Set a closing and a cut-off based on a date or on the number of versions. Update this as you go.
- There are so many transactions to support and there is limited time for the prototype. What do you do? After critical transactions have been supported, you can defer these until you have an initial production version. They can be added then. How many transactions should be in the prototype? Start with the 80–20 rule; that is, 20% of the transactions usually account for 80% of the volume.

REDUCED SCHEDULE AND COST APPROACH

Get many users involved in the definitions of transactions. Do not have the work fall on the shoulders of a few people. Arrange for meetings where they can walk through what they have done. Use the pilot and the enthusiasm generated to get more people involved. From experience, the detailed definition of the transactions and business rules can be a major bottleneck for development.

Another bottleneck is the area of interfaces. If you can, get the interface design and development started during the prototyping. More is said about interfaces in Chapter 6.

For the development of the prototype, first establish the database and tables. If you are going to reuse the prototype in production, then consider establishing these on the production database management system as opposed to a PC database. This will cut down on time later. To cut costs, carefully review each of the initial programs written for the prototype. Focus at the start on what code and libraries exist so that you can cut down on development. Emphasize getting the programs up and running as opposed to programming elegance. To better organize the work of the programmers, line up their assignments in a detailed way. This ensures that there is a better understanding of the work.

With respect to the pilot, conduct the demonstrations in the business departments. After the first pilot demonstration, suggest that the business department take over arranging for the pilot. This will free up your time for other work.

EXAMPLES

For the Roberts Agency, an intranet system was designed and developed for bus drivers to fill out forms and track lessons learned, lost and found items, and other daily functions. The system was intended to be used by drivers in their homes. Therefore, the method was to use intranet and not clientserver. The prototype included separate menus for each major function. Linkage among menus was handled by web page links. The first prototype served as the basis for the working relationship between business staff and IT programmers. Linkage with electronic mail was established early and used as an aid to routing transactions.

During the development, more than 15 new applications were requested by bus drivers during the pilot demonstrations. These were catalogued to be prioritized and analyzed later. They were not implemented. The entire focus was on transactions. There were more than 40 transactions. Of these, 10 constituted more than 85% percent of the volume; 15 transactions were only used once or twice per year and were deferred until later.

The pilot experience was very successful. There was so much enthusiasm that additional demonstrations were held at the union headquarters. An effort was made to define the benefits of the new process. Here is what was discovered through the pilot and by projecting the results into production. The benefits exceeded the initial estimates.

- The bus drivers could save 5 to 10 minutes in filling out forms at the bus base by filling them out at home. This amounted to substantial savings when you consider that there were 900 bus drivers and that several forms were completed each week by every driver.
- The workflow indicated that the review of the transactions could be handled at one bus base. In the old process, this was done at each base. With three bases, this freed up two employees.
- Bus drivers could log on and see the status of their transactions and requests. This reduced telephone calls to the base to determine status and saved the staff more than 5 hours per week.
- There was virtually no training because more than 90% of the drivers had PCs at home with internet access. Of the remaining drivers, most could access the system from a friend's house or at the local library.
- Bus drivers felt more in control of their work, thus morale and the drivers' relationships with the agency were improved.

The new system had to interface to 25-year-old legacy system. The data elements in the new system were much more extensive than those in the legacy system. This created a need for a program to extract from the database and establish an intermediate file that matched the format of the legacy system. Due to concerns about this interface, work on this program started during the prototype.

LESSONS LEARNED

- Use the transaction information as a basis for negotiating with department managers and staff on when the system should go live. If you wait until all infrequent transactions are programmed, then implementation will be deferred too long. Pose it to the department as a trade-off.
- The stress on configuration control of the prototype cannot be underestimated. The project leader should be very actively involved in specification, review, and support of the reuse of code.
- Constantly enforce change control; however, you should maintain a list of potential changes, who proposed them, and their potential benefits.

SUMMARY

The prototype approach has been shown to work because it provides business staff with tangible evidence of how the new system will work. It is much more effective than presenting a thick document in a three-ring binder. However, you must be careful not to raise expectations and to think that the prototype can be used in production.

Overall, for development, the basic recommendation is that you use as few methods and tools as possible. Keep it simple. Keep it effective. Adopt an object-oriented approach around transactions. You will end up with a simpler design and simpler code.

WHAT TO DO NEXT

- Review the existing methods and tools and see what could be employed in developing prototypes. What additional tools would have to be acquired?
- Review the literature on prototyping on the web and see if you can find case examples of prototyping that would fit your situation.
- In thinking about prototyping, what barriers might you encounter? Consider both the business department and IT.

Chapter 6

Perform System Development and Integration

INTRODUCTION

In this chapter, we take a moment and consider the current status of development. In Chapter 5, the requirements including the interfaces, were defined, and the prototype and pilot were completed. The user interfaces, menus, screens, databases, and some business rules are finished and stable. However, business staff are still defining transactions and business rules. The tasks of conversion, documentation, and training are addressed in Chapter. Here the focus is on the extensive work required to develop, integrate, and test the system. The specific activities include the following:

- System and program design
- Programming and unit testing
- Building and testing interfaces
- Integration of programs and testing
- Testing at a system level

New methods and technology have streamlined some of this effort. Because of the extensive work with business staff in prototyping, there is less of a need to produce design documents for review. After all, business staff are defining the business rules and transactions and they have seen and reviewed the screens, navigations among screens, the database, reports, and other parts of input and output.

There are now more advanced programming tools than existed before. Integrated programming tools and a programming environment have facilitated the programming work in terms of editing, debugging, documenting, and testing an individual program. There are also tools that generate test data and analyze test results. Programming libraries are also more extensive.

Nevertheless, there are still major areas of concern and risk. Integration to legacy systems requires customized application program interfaces (APIs). Integrated test tools are still limited. Batch programs to perform updates, backup, and so on must be written. Finally, the team must program the business rules.

The development approach that is recommended here is a combination of proven techniques:

- There must be a push to *reuse* existing programs and use program libraries. This will cut down on the amount of new code required.
- Gathering of detailed business rules, design, and development are carried out in parallel with development. This helps to reduce the elapsed time. This principle is based on that of *concurrent engineering.*
- Detailed project management, *configuration management,* and *change control* are major foci of development.
- More active involvement by business staff in the project leads to greater participation and commitment as well as improved communications between the business and information technology (IT).
- Using *object-oriented methods,* the development is based on building smaller modules and programs. This facilitates tracking and control as well as reuse.
- Minimize documentation and focus documentation and tools on programming, integration, and testing.

COMPUTER-AIDED SOFTWARE ENGINEERING

Computer-Aided Software Engineering (CASE) is a name given to methods and tools that attempt to automate the analysis, design, and programming of systems. A few of the principles behind CASE are as follows:

- If methods and techniques can be automated for analysts and programmers, then there will be substantial productivity gains.
- Much of the design work can be automated. Once the design is automated, it can be modified, reviewed, and updated much more easily than a paper-based system.

CASE tools emerged in the 1980s with software such as Excelerator, Visible Analyst, and others, which were PC-based packages that contained a number of development tools. These tools were then linked together to share data. Texas Instruments IEF (Integrated Engineering Facility) attempted to go further and generate computer code or pseudocode. Ideally, you could enter requirements in one end and evaluate them after they were analyzed, structured, and output. The information then served as input into design. Finally, you could take the design and use it to generate code. The dream was to have a software factory.

The benefits claimed by firms who offered CASE tools included the following:

- Improved user–systems analyst communications
- Ability to maintain and modify requirements and design
- Analysis capabilities to examine the completeness and consistency of designs
- Improve productivity of the designer and programmer

CASE tools have been grouped into upper CASE and lower CASE categories. Upper CASE refers to tools that support the creation and modification of the system's design, whereas lower CASE tools include those that generate code.

CASE tools have some drawbacks. They are expensive and time consuming to learn and use. Many of the CASE tools demand that the requirements are known. Most tools do not create designs; instead, they evaluate requirements and design for consistency. CASE tools have been criticized because they consume too many resources in projects with limited time and money. CASE tools often do not interface with each other; thus, additional effort would be required to take the output from one tool and input it into another tool. Many believe that the effort is better spent on prototyping and piloting.

CASE methods spurred software firms to include CASE capabilities in their software tools. As a result, many CASE components can be found in the programming environments for C^{++}, Java, and other modern languages. With the use of prototyping and the emergence of object-oriented methods, the need for other CASE tools diminished. The use of CASE methods has also entered into database management systems and fourth-generation languages that now offer some of the same CASE capabilities.

A number of CASE tools have fallen into disuse because they did not live up to their promise. One company won a contract to employ an integrated CASE tool to develop a major system for a state's motor vehicle department. It was a total failure. Millions of dollars were lost. CASE methods are, however, valuable and have now become embedded directly in development and programming tools as opposed to separate, stand-alone CASE tools.

SYSTEM DESIGN

Your system design and the techniques you employ must reflect certain facts:

- Requirements for the software will be changing and will never be complete.
- The software that you develop from the design will tend to live a long time.
- The system will be applied to situations that were unanticipated when development was done.

The objective of system design is to produce a design of a system and its programs that is complete and modular, usable by programmers, carries out the requirements, and will result in a maintainable system. The goal of an efficient design has been replaced by an effective design due to improvements in hardware, system software, and networking. The use of modular design results in many, smaller programs that then share data through standard interfaces, such as ODBC (Open Database Connectivity), OLE (Object Linking Embedding), Active X, and so on. The general design effort has been reduced in scope and has been affected by the following factors:

- The graphical user interface (GUI) is often now set in the prototype, eliminating this from the design.
- Design of programs to generate reports is reduced because many reports are now replaced by producing output files for a spreadsheet or an updated database where report generation tools can be employed.
- Certain functions that had to be part of design and programming have now been replaced by the use of programming libraries. Programming libraries are extensive catalogs of canned, tested software that are used as building blocks in development.

Here are some design guidelines:

- The design is top down and individual programs are designed around specific business transactions or batch functions.
- Programs are grouped into classes where each class performs a general type of function, such as data entry, updating, and reporting.
- Considerable time and effort go into the design of the first program in a class. This design is then copied and reused to design other programs in the same class.
- Design today requires a collaborative approach where the system designers share information extensively.
- Emphasis in the design is based on providing specific capabilities to support individual transactions and functions.

A useful technique in design is to get business and IT staff together to hash out business rules and elements of the system design. A formal approach for doing this is *Joint Application Design* (JAD). There are numerous variations on how this is done. The purpose of JAD is to reduce the time for interviews and data collection as well as to deal with issues in a group. The general method is to have an organizer or moderator who conducts the session and observers who attend the session and offer technical explanations and support. There is extensive preparation for the session. The organizer interviews the participants of the upcoming session and identifies issues in business rules, workflow, and design. These are then considered one by one in the JAD session. There must be someone who will document the results of the JAD session for later review.

To support JAD sessions, the culture of the organization must support joint problem solving. If there is open hostility among different departments, the effectiveness of the JAD is affected. The workload of the participants allows them to be absent at an off-site location for a 4-hour period or longer.

There are some drawbacks to JAD. One is that a substantial block of time is consumed for 15–20 people to participate. A second drawback is that the topics may be too politically sensitive for the group to address. They must be dealt with by upper management. A JAD can also polarize different segments of the group around opposing sides of an issue. Some issues may not surface due to politics. Thus, there must be detailed preparation and follow up to the JAD if it is to be successful.

How do you document the design? There have been more than 75 techniques developed. Because the data base and the GUI have been defined, the program design concentrates on the business logic. Some variations of (Input Process Output (IPO)) have proven useful. Flowcharts and decision tables tend to be too cumbersome and difficult to maintain; however, because you are doing a modular design with many smaller programs, this is less of an issue. Also, you will probably not use these techniques since the detailed design of programs performing similar transactions in a class are so similar. They will be clones of the first design.

PROGRAMMING AND UNIT TESTING

PROGRAMMING TOOLS

There are now sophisticated suites of integrated programming tools that provide an environment for programming and unit testing. Capabilities include the following:

- Wizards to generate initial code
- Canned drag-and-drop components
- Shared reuse of modules across multiple languages
- Generation of HTML code
- Visual models that support program design
- Linkage to database management systems
- Interpreters and compilers
- Test aids for debugging and tracing program execution
- Generation of test data
- Word processing editors
- Data dictionaries
- Documentation aids
- Decision table generators
- File compare utilities

- Online debugging that supports checkpointing, restarts, and modifications
- Source code formatters
- Syntax checkers
- Cross-reference listers
- Debugging compilers

These suites implement much of what was envisioned for CASE programming workbench tools.

EXAMPLE OF CHANGING TOOLS-INTRANET DEVELOPMENT

To show you how fast the software development is changing, we consider the evolution of intranet development. Originally, when the World Wide Web was developed, there were only HTML (Hypertext Markup Language) pages. When you input the URL address into the web browser, the server would find the appropriate web page and download it to your browser. Your browser read the HTML tags and created the picture. Communications were basically one way; however, developers pushed for two-way communications. The first effort was to allow for forms, which are HTML tags that allow for edits, checkboxes, and radio buttons. Then the Common Gateway Interface (CGI) allowed the web server to communicate with an application when the form was sent back. Programmers used the language Perl to save the form to a database. This was both a development and performance nightmare. If there were six users who worked with the same form, then the web server had to have six copies of the CGI running at the same time. It is no wonder that performance suffered.

The next step was server APIs. These allow for only one program to run but require complex coding, which was not easy to work with.

The next generation was support for more English-oriented languages such as Java and Visual Basic. As an example, Microsoft developed Active Server Pages (ASP) so that a program code could be written within the HTML page. A simple scripting language such as VBScripts or JavaScripts could be employed. These are stripped-down versions of Visual Basic and Java, respectively. At the web server, you can link the code to a database using ODBC.

With ASP, no compiling or complex interfaces are required. Moreover, performance is improved because the code is executed on the server. There is additional security because the code is not downloaded to the browser—only the HTML page is. Because only HTML is sent back, any browser can be employed.

Tools such as ASP continue to evolve. They represent a significant advance in programming for online applications in general and are now replacing more complex clientserver applications. Fourth-generation languages such as Oracle have also implemented intranet development tools into their toolkits.

UNIT TESTING

Unit testing involves undertaking the following steps:

- Walkthrough of the source code to determine if the design and requirements were met
- Execution of simple tests on the program to verify the program structure
- Testing of the program using valid and invalid input data and review of results

Unit testing of the first program in a class is extremely important because it serves as a model for the other programs in the classes to follow. These classes will require less review because they are based on this program. Specific testing can include the following:

- Dynamic analysis of program code to find errors through manual review and execution.
- Static analysis to determine any omissions. This can be accomplished through data flow analysis and inspections. Errors in data and type declarations, mathematical expressions, and interfaces can be detected.
- Coverage-based testing. This is the testing and review of a program to see that each statement in the module or program is accessed.

Building smaller programs generally reduces this effort. For existing legacy system COBOL programs, these tests are difficult if not impossible to undertake due to the size and complexity of the program code. Imagine, if you will, a program that is thousands of lines long and has been built and maintained by eight people for more than 20 years and you can see the situation.

As you test you will find errors. You will then work to debug the program and fix any errors. With simpler modular programs, this is usually quite easy. However, with more complex programs, you can benefit from debugging aids. These include error checking software, automated flowcharting of the program, and tracing of statements and values of variables during the execution of the program.

TESTING PREPARATION AND TEST CATEGORIES

SYSTEM TEST PLAN

A *system test plan* provides everyone with a roadmap for testing. The size and detail of the test plan depends on the complexity of the system and its environment. An outline of the test plan is as follows:

- Introduction—purpose of the testing, technical architecture of the system, scope, hardware and software, assumptions made in testing
- Requirements
  - Functions—requirements for all functions (entry, update, inquiry, etc.) to be tested
  - Integration—requirements for interface testing
  - Design—requirements for user interface (this has usually been handled through the prototype)
  - Test strategy—how purpose of testing will be met by the type of test, test cases, completion criteria, tools
- Project plan for testing
- End products of testing

TEST PREPARATION

To do testing, you must complete a series of steps. These are as follows:

- Design and build test suites. A *test suite* defines a specific test to be carried out, the purpose of the test, the input for the test, and the anticipated test results.
- Determine how testing will be done. Will testing be done manually with automated testing aids and tools?
- Decide on how test results will be reviewed and used.
- Establish a testing environment and repository for the test data and test cases.

There are several types of test data:

- *Live data.* These data are from the current production system that is used for load and stress testing as well as for running parallel tests between the old and new system.
- *Created data.* These are test cases that are either manually constructed on a field-by-field basis or generated by an automated test data generator.
- *Modified live data.* Live data are modified to carry out specific tests.

All three types of data should be used in testing. For example, created data are useful earlier in testing, modified live data are of value when you are attempting to test various extremes, and live data are needed for comprehensive testing.

When an error is discovered, it must be documented in a database of errors. The following information is needed to assist the programmer:

- Error identifier number
- Who is reporting the error
- Date reported

- Version of software tested
- System and programs tested
- Severity of error—system does not function, system functions but does not give correct result, minor error that does not have an impact on the results
- Symptoms of error
- Input that gave rise to error
- Anticipated result of test
- Comments

Errors can be due to a number of factors. These include the following:

- *Design errors.* These include data design and communications errors. This category is lessened in modular development.
- *Programming errors.* These include syntax errors and design misinterpretations.
- *Clerical errors.* An example is an input error.
- *Debugging errors.* The debugging tool was improperly used.
- *Testing errors.* The system was not properly tested.

When the programmer works on an error, he or she removes the program from the development library and downloads it to his or her PC. There he modifies the code and performs unit testing. Upon completion, the program is returned to the library. It is critical that there be an audit trail of the changes made to a program. This is especially true if there are several programmers who are going to work on the program; each has to know what the other did. The programming environment tracks changes to the program. The programmer needs to add documentation that includes the following:

- The error and its identifier
- Name of the programmer
- Date of change
- Test data used
- Explanation of the cause of the error
- Changes made to the program to fix the error

Although at times there has been an effort to reduce documentation, this documentation is critical and its development must be enforced and the work must be reviewed. This is a key role for the Quality Assurance function.

There are some useful guidelines for debugging:

- The programmer should concentrate on one error at a time.
- Embed debugging aids within the program during initial development and disable them. During debugging, these can be activated.
- Give attention to data handling by a program instead of just concentrating on program logic.

- Use the best debugging tools available. This is obvious, but the programmers must agree on the same tool. Share experiences in using the tool.
- Keep an eye out for programs that are complex and large. Consider breaking these into smaller programs.

TYPES OF TESTING

Integration testing is discussed in the next section. You also want to test the system to see how it performs under a load of work (*load testing*) and under stress conditions (*stress testing*). You will also need to perform testing of the entire system (*system testing*). After system testing, it is time for the business staff to test the system. Although they have done testing during the prototype, they do in-depth testing on transactions to validate that business rules have been correctly implemented (*acceptance testing*). In the past, these tests were performed in sequence. Today, with modular development, users are doing acceptance testing much earlier for transactions that have been completed.

INTEGRATION OF PROGRAMS

When you integrate two programs, you establish a link between them. This is referred to as *coupling*. The looser the link and tie between two programs, the better. This is because if one program controls the internal workings of another, maintenance, testing, and completion are much more complex. You have to work on and test both programs. The weakest form of coupling is data coupling. In data coupling, data are passed from one program to another. This is the best form of data coupling.

When you begin integration, the programs reside in a library or repository. You will copy these from the library and link them together. Then you move them into a test environment. Here you will apply the test data and test suites. You continue in this mode. Because you are doing modular development, there is more integration (programs are more numerous and smaller). However, if you have standard data interfaces, then this will ease the integration work.

When the entire system is completed, the programs move into a production library. For purposes of configuration management, there should be three separate environments: development, test, and production. *Development* is where the programs write and maintain the code and perform unit testing. *Test* is where integrated testing occurs.

INTEGRATION TESTING

When you are ready to do integration testing, the individual programs that you are to integrate have already been unit tested. The objective of integration testing is to ensure that each program performs correctly as part of the interface and that the interfaces themselves are correct. To do integration testing, you combine the programs in stages. At each stage, you add a program to the existing structure. You then carry out tests for the combined programs.

There are several ways to do integration testing for larger systems. One is *bottom up testing*. Here low-level programs are integrated first. You work your way up from the bottom to programs of increasing complexity. To do bottom up testing, you have to create data to feed these programs (called *drivers*). These programs require one-time effort and consume resources. High-level functions are not tested until the end.

In *top down testing,* you test higher-level programs first; then you proceed downward. To do top down testing, you have to develop and implement stubs or dummy routines that respond with data when a high-level program calls a lower-level program. As you work your way down, you replace the stub by the program code. This method tests the high-level interfaces first and is a common approach in object-oriented development where there are standard interfaces and where programs are more modular. For complex systems, top down testing does not test low-level functions until late in the development.

SYSTEM TESTING

System testing occurs after integration testing to ensure that the entire set of programs is functioning properly. In the modular development approach, system testing is simpler due to the modularity. Here are some steps in doing system testing:

- Use the requirements and specifications to construct test cases.
- Determine the expected outcomes of the tests prior to testing.
- Start with the simpler test cases and build to the more complex test cases.
- There are three stages in testing: test normal conditions and data test extremes for input variables and volume, and test exceptions to see that the system rejects invalid input.

WHAT CAN GO WRONG?

IT management may decide to adopt a new technology, method, or tool. They launch and endorse it with enthusiasm. This is followed up by training. Yet, the approach falls flat on its face. Why did this happen? Here are some things you must do to be successful:

- Determine exactly how the method or tool will fit within the development environment. Decide which projects and work are most suited to the tool.
- Undertake an initial project using the method or tool to gain lessons learned before pitching it to everyone.
- After the initial project, gather lessons learned and create guidelines for effective use. Also, identify an expert who can assist in subsequent applications.

The change control method was not fully endorsed or understood by management. Business conditions and situations change, and new priorities are issued. The project leader of the development team and the team do not question this. Instead, they attempt to adapt to the change. The project plan is not updated and thus becomes unrepresentative of the real project. Failure is on the horizon. When there is a business change, it is important to show management that there is a trade-off. The direction can change but so will the schedule, risks, and cost.

REDUCED SCHEDULE AND COST APPROACH

Following are some guidelines that have been employed by several firms to reduce the schedule and cost of development. These include the following:

- Enforcement of the reuse of software and use of standard programming libraries. This requires a library of programs to be established. Programs must be clearly documented and tested. There must be a role defined for someone to facilitate the use of the library, and the use of the library must be enforced. One way to do this is to measure the extent of intended reuse during the design activities.
- Develop and integrate the software as you develop. This requires analysis at the front end to structure and plan the work so that subsystems can be established and tested without waiting until the end of the project.
- Ensure that there is accountability for each programmer as to what specific programs he or she is developing. Avoid any general assignments. Use configuration management to track the code and assignments as well as project management.

EXAMPLES

As businesses become more dependent on technology, the impact of defects and errors in system code grows exponentially. The costs of software errors cannot be underestimated. For example, several major airlines lose $20,000 in nonrecoverable income each minute when their airline reservation systems go down. A manufacturing firm estimates that they lose $50,000 per minute when their production line goes down. A credit card operation estimates that their losses exceed $150,000 per minute when the credit card authorization system fails. Finally, consider automotive manufacturers who embed technology and chips in their vehicles. A recall for an embedded chip can cost $100 per car.

A major theme in this chapter has been software reuse. Here are some examples:

- Toshiba has implemented a reuse division. As a result of corporate support, reuse rose to 32%. Breakeven for reuse occurred after three uses.
- Fujitsu established a software support library and mandated that it be employed. As a result of reuse, there was a 20% improvement in project schedules.
- GTE Data Services established a reuse program with financial incentives to IT staff. The first year reuse rate was 14%, with a savings of $1.5 million. The company projected a rate of 50% reuse in the future.
- Celsius Tech Systems achieved a 70% reuse rate, with substantial cost savings.
- Raytheon implemented a reuse program in a division for its business systems involving COBOL programs. With 60% reuse there was a 50% net productivity improvement during a 6-year period.

LESSONS LEARNED

In development, there are four types of risk:

- *Business*. Business risk occurs when the business environment and business process are undergoing change. For example, you may start the development assuming that the company will use a centralized customer service group. During the development, management decides to outsource customer service to three different firms in different regions. To head off business risk, you must stay in touch with the business department and management to detect potential changes. Then you can alert management to the potential risk before decisions are finalized.
- *Technical*. Here the risk is that the system will not work as required. Technical risks can be minimized by using established technologies, methods, and tools from the start; by exercising technical change control; and by identifying and dealing with issues quickly.
- *Schedule*. This is the risk of the work falling behind. By taking the modular and concurrent development approach, you help to reduce this risk. You also

minimize the schedule risk by involving the business staff heavily in the project. This reduces exposure if there is a slippage.

• *Cost.* Development projects have several categories of costs. There is the cost of the technology infrastructure, which can be controlled by fixing the technology, methods, and tools early on and by change control. There is also the cost of the software database management systems and other software to support development. This can be controlled by analysis and definition at the start. The third cost is labor costs, where you rely on concurrence, reuse, and modular development.

SUMMARY

When performing implementation, you give the highest priority for tools and aids to programming, integration, and testing. These are also the areas where you want to be the strongest in enforcement of standards and methods. In addition, configuration management is a critical tool that can be used to help in monitoring and controlling work and changes.

WHAT TO DO NEXT

- Review the current design and programming methods and tools. Is there a standard set of tools or can programmers select what they use? Are there standards for the use of the tools? Are there examples of correct practices? Are standards enforced? Is program code reviewed? Are the tools integrated?
- Consider developing a project plan template for all new development projects. Along with the template, you should gather and establish a database of common issues that have been encountered in development. This can reduce the risk, as well as the time, in getting a project started.

Chapter 7

Conversion, Procedures, and Training

INTRODUCTION

This chapter addresses three topics that typically do not receive enough attention but that can have a major negative impact on implementation if not carried, out correctly. *Data conversion* is the transfer and manipulation of data from the current process into the new system and process. *Procedures* refer to both operations methods and procedures for using the new system. *Training* includes business department training. The material in this chapter also applies to software packages and knowledge management.

DATA CONVERSION ISSUES

What does "doing conversion right" mean? Data conversion ensures that the new system and business process will have data that are as correct and complete as possible for production. This statement means more than it implies. Data in the current system typically have some or all the following problems and shortcomings:

- Data are not compatible with the specification of information in the new system.
- New data elements have been added in the new system and are not to be found in the current system.

- Some data in the current system, although defined, may be missing.
- The data for the new system may have to be drawn from multiple existing systems. These existing systems are not compatible with each other, making your work harder and creating inconsistencies.
- There may be problems with data quality and completeness in the current system.

For Atlas Bank, the existing checking account system had a field for telephone numbers. When the new collection system for credit cards was implemented, it was believed that the telephone number in the checking account system could be used. Programs were written to capture this field. No one checked out the data. It turned out that less than 15% of the telephone numbers were accurate. The impact of this very unpleasant surprise was that the telephone numbers had to be captured manually and entered by hand. This delayed implementation by 3 months.

Turning from the existing data, basic questions such as the following remain to be answered:

- Where will the missing data be found and entered?
- How will new data be captured and entered?
- How will data in the new database be validated?
- What are the approach, timing, and testing of data conversion programs to establish data in the new system?
- Will there have to be multiple conversions performed—the first for testing and training and then another for production?
- Can initial production be performed without all the data?
- What happens to the data in the current systems? Will there be a requirement for ongoing interfaces from the existing systems into the new system?

Generally, when you do conversion, you first write and test conversion programs. At a certain point, you run the conversion program to establish a test database for the new system. This is employed for testing the new system, doing user training, and developing procedures and training materials. Just before the new system is going to go live in production, you perform a final conversion of the data that contains the latest data in the current system.

DOCUMENTATION AND PROCEDURES ISSUES

Depending on the system, there can be a range of documentation, including the following:

- *Operations and production procedures to run the system.* These include backup, recovery, and restart procedures as well as instructions for normal batch and online processing.
- *Network and security procedures.* Network recovery, troubleshooting, access control, diagnostics, and normal network monitoring and operation are included.
- *Business process procedures.* These encompass procedures and policies for performing the business process overall.
- *User procedures.* Traditionally, this consists of a manual on how to use the new system.

In addition to these written procedures, you can have online help. You can also distribute the documentation via web pages by establishing a web site. This eliminates distributing updates of the materials via paper.

What are some of the critical decisions that you must make?

- How much time and effort can be budgeted for generating documentation?
- How will documentation be maintained and updated?
- Who will create the documentation?
- What will be the form, format, and structure of the documentation?
- How will you measure the effectiveness of the documentation?

Keep in mind that experience paints a grim picture of documentation that is generated, but never used or maintained.

TRAINING-RELATED ISSUES

Training includes the generation of training materials, the initial training of business staff members, and recurring training of new and existing business staff in regard to updates. In the past, the training responsibilities were split among IT and business departments. IT would generate training materials for the system and then train users in the system. The business department would then train the staff in the business process and generate any necessary training materials. This approach is seriously flawed due to the following:

- There is no provision for how new employees in the business department will be trained in the system. IT staff have rarely returned for follow-up training.
- Users become confused by having two separate training courses for the system and process, especially when the system is contained in the process.
- IT-generated materials are often hard to understand for end users. These materials may contain technical jargon and assume that readers have more knowledge of IT than they do.

STEPS IN DATA CONVERSION

These steps can be followed for data conversion:

- Step 1: Review the current data and data requirements of the new system.
- Step 2: Decide on the approach to conversion.
- Step 3: Develop a detailed project plan for the conversion.
- Step 4: Write and test conversion programs.
- Step 5: Establish a manual conversion method.
- Step 6: Perform the initial conversion and test the results.
- Step 7: Perform the final conversion.

Although these steps appear to be sequential, there is substantial overlap because you will typically find out more about data problems and issues as you go.

STEP 1: REVIEW THE CURRENT DATA AND DATA REQUIREMENTS OF THE NEW SYSTEM

Given the time that these tasks require, you should proceed in parallel with both. In terms of the new system, determine the answer to the following question, "What are the minimum data requirements to go live with the new system?" Remember that the new system has more fields than the old. Although it might be nice to populate the new database, your focus is on implementing the system quickly. You may also find out in production that some data elements are not needed.

For the current data, you can prepare a data mapping that relates the data elements between the old and the new systems. You could define the following table. Figure 7.1 lists each data element that is in both systems in the first column. The second and third columns give the name and characteristics of the data element in the old and new systems, respectively. The fourth column explains the difference, whereas the last column discusses any conversion issues.

This analysis is more work than you would believe because data elements can have the same name but have very different meanings. If necessary, the analysis must be extended to multiple source systems if these are going to feed into the

Data element	Current system	New system	Difference	Conversion comments

Figure 7.1   Data mapping.

new system. This data analysis helps in the establishment of interfaces and in doing integration.

STEP 2: DECIDE ON THE APPROACH TO CONVERSION

Here are some alternatives to conversion that you should consider:

- Do nothing. Do not convert. Add data in the new system as you go. This is possible in some systems in which you print the history from the current system and then use image, paper, or microfilm to analyze it.
- Convert as much as possible from the current system using conversion programs and correct the data later.
- Convert as much as possible from the current system using conversion programs and correct the data now through manual effort and research.
- Selectively convert only critical fields that are valid and then update the missing data based on importance.

Many people try to adopt the second alternative. However, this approach attempts to assume that most of the current data are okay. The preferred alternative is the last one, but this requires more analysis.

STEP 3: DEVELOP A DETAILED PROJECT PLAN FOR THE CONVERSION

Why do you need a project plan? This seems straightforward. Programming interfaces and conversion programs take time and must be coordinated with the overall implementation schedule. If you are going to review paper files and correct data after conversion, then this will be a sizeable manual effort that must be planned. The following list provides some of the major tasks in the plan:

- Identification of interfacing systems
- Analysis of current and interfacing systems data
- Analysis of data elements in new system
- Minimum data element requirements to go live in production
- Comparative analysis of data elements between current and new systems
- Determination of conversion approach
- Conversion programs
  - Specifications
  - Design
  - Programming and testing
- Run of conversion program to get test data in new system

- Sizing of manual conversion requirements
- Definition of procedures for manual conversion
- Staffing for conversion effort
- Manual conversion effort
- Evaluation of converted data

STEP 4: WRITE AND TEST CONVERSION PROGRAMS

These programs are typically COBOL programs that extract from legacy systems. The most frequently encountered problems tend to be as follows:

- There are misunderstandings about data elements so that individual data elements are converted wrong.
- After the initial programming, maintenance on the current systems leads to database changes. These changes are not communicated to the conversion programmer. When the conversion program runs again, it blows up or moves the wrong data.
- There is a lack of effort made to evaluate the data that were converted.

STEP 5: ESTABLISH A MANUAL CONVERSION METHOD

A manual conversion effort begins with identifying the data that are to be captured along with the source of the data. Examples of data sources are microfilm files, paper files, outside sources (e.g., telephone information), and reports. You should identify what furniture, computer equipment, supplies, and other equipment will be needed. Then you can begin to define a new workflow to capture and enter the data. You also have to define how the data will be verified and checked.

After you have defined the workflow, you can determine the staffing requirements necessary to complete the work within the schedule and define the skill levels needed. Next, you will have to develop training for the people. Once the conversion effort starts, identify how feedback from verification of the data is sent back into the conversion.

STEP 6: PERFORM THE INITIAL CONVERSION AND TEST THE RESULTS

Whether by programming or manually, you will want to evaluate the results of the initial conversion. Here are some suggestions:

- Identify critical data elements and select a sizeable sample of converted records. Review these specific fields in the converted records of the database. You may want to generate a report to get this information.
- Take a smaller sample and review the other, noncritical fields to see what happened in the conversion.
- When checking the data, you are looking at the format of the data, alignment of the data in the field, and the actual value of the data, as well as for any spurious data.

STEP 7: PERFORM THE FINAL CONVERSION

After doing the conversion, you should still check the data. Continue to spot check records. What do you do if you find a problem that was undetected? You cannot stop the production system; it is too late for that. The best approach is to analyze the data to see how bad the error is and then write a specific program to correct the individual fields. Do this as a batch job when the system is not being accessed.

STEPS IN DOCUMENTATION AND PROCEDURES

The steps in documentation and procedures are as follows:

- Step 1: Determine documentation requirements.
- Step 2: Define the documentation approach.
- Step 3: Develop system operating, network, and security procedures.
- Step 4: Develop user and business process procedures.
- Step 5: Evaluate, test, and revise the documentation and procedures.

STEP 1: DETERMINE DOCUMENTATION REQUIREMENTS

What can you afford? What will really be used? How much can you maintain and update? These are some of the questions to answer here. A lesson learned is to determine the documentation needed by considering self-interest. That is, answer the question, "What do people want to know to help themselves?" The answer here can eliminate a lot of general description documentation and get you down to the "nitty gritty" of the business process transactions.

Another perspective is based on risk and exposure to problems. You want to concentrate on areas where documentation can prevent errors. For example, suppose that security is not a major concern because the system is internal and because the data are not confidential. Then because there is limited risk here, you might spend more time in other areas, such as exception processing.

STEP 2: DEFINE THE DOCUMENTATION APPROACH

Who do you want to be involved in generating and reviewing the documentation? What documents from other systems can be employed as models so that you do not have to start from nothing? Because your time is very limited, aim at approaches that do not require you to reinvent the wheel. It is good to be creative in implementing a system, but you want to avoid this with respect to documentation.

STEP 3: DEVELOP SYSTEM OPERATING, NETWORK, AND SECURITY PROCEDURES

The advice here is to follow what has been established for other systems. Also, interview the current computer and communications operations staff to see what suggestions they have that would make it easier for them. See if you can get them involved in the review of the procedures as well.

STEP 4: DEVELOP USER AND BUSINESS PROCESS PROCEDURES

Here you seek a combined set of procedures for working with the system and the business process. If you keep these separate, then you risk confusion. How should the procedures be developed? Keep in mind that the knowledge lies in both IT and the business department. This supports a joint approach. Here is a list of actions:

- Action 1: Establish a team to develop the procedures from both IT and the business department. Define the roles of what people will do in terms of definition of procedures, development, and review.
- Action 2: IT trains the business staff in the system. The business staff indicate how the process interacts with the system.
- Action 3: The procedures are defined. This includes an outline, how the procedures will be used, the level of the audience, and how the procedures will be updated.
- Action 4: Develop a sample of the procedures and test it out with people from the real audience. Based on feedback, refine the sample.
- Action 5: Divide the work for the procedures and start developing sections in parallel.

STEP 5: EVALUATE, TEST, AND REVISE THE DOCUMENTATION AND PROCEDURES

After the procedures have been developed and tested for accuracy and usability, you must still test them in terms of the business process. This goes beyond the normal evaluation of procedures. Here you take the procedures to the process. Have individuals describe exception transactions and odd situations. Using only the procedures, see if you can determine what to do with these. Do not try to improvise because the staff will not be able to do so when they get the procedures. This approach is also useful in testing the technical operating procedures as well.

In addition, you must determine the final form of the procedures and how they will be distributed and controlled. Here are some comments on alternatives:

- *Online help.* This is not as accessible as a manual. Thus, this alternative is best employed for specific transactions rather than general help.
- *Paper manuals.* These are easy to access, but updating them is uncertain.
- *Use of laminated cards and "cheat sheets", along with the manual as a backup.* This is probably the best traditional approach.
- *Web-based documentation.* This is a good idea, but many people are not comfortable with this or a personal computer (PC) may not be nearby when they need to refer to it. Due to time pressure, a person will then improvise rather than read the procedure online.

For written procedures, consider using the time-tested approach of a playscript. Here the page is divided vertically in half. On the left is the person or computer; on the right are bullet points and steps as to what each does. This format follows that of ancient Greek plays—a time-proven approach.

STEPS IN TRAINING AND USING TRAINING MATERIALS

Following are the steps for training and using training materials:

- Step 1: Define requirements, roles, and approaches for training.
- Step 2: Prepare training materials.
- Step 3: Evaluate, test, and revise training materials.
- Step 4: Set up for training.
- Step 5: Conduct training.

STEP 1: DEFINE REQUIREMENTS, ROLES, AND APPROACHES FOR TRAINING

Begin by answering some basic questions about the audience:

- How many people must be trained?
- What are the average and low skill levels that you can assume?
- How formal is the current business process in terms of procedures?
- What is the level of staff turnover?
- How is work assigned to the staff? Do certain people only perform specific sets of transactions?
- What types of training have they had previously in the business process?

Answering these questions leads you to identifying what to do to answer each of the following:

- Who should do the training?
- Will the training target specific groups of people or be general in nature?
- What types of training materials are needed, and what will their purpose be?
- What will the format and structure of the training be (number of sessions, size of room, equipment needed, etc.)?
- How long will the training sessions last? How many sessions will there be?
- What will the trainees do to prove that they did learn?
- How will the training be measured and monitored?

STEP 2: PREPARE TRAINING MATERIALS

Try to turn this from being tedious work into something more fun. Develop a single training module. An example might be materials for a group of transactions. Take these over to the staff to be trained and have them review and comment on them. While you are there, talk to supervisors to see if there is someone who likes to work with other people and who might be available to assist with the development of materials. Develop materials from a detailed outline. Keep a plan so that you can track your progress.

STEP 3: EVALUATE, TEST, AND REVISE TRAINING MATERIALS

Review the materials as they are developed. First, read the materials for consistency and accuracy. A second pass can concentrate on style and simplicity. The best test for the materials is to have people follow the procedures by doing the process and using the system with only the materials and no in-person instruction. What are you looking for?

- Are the materials clear or ambiguous?
- Is it easy to move among the sections of the materials?
- Are there missing steps in the detail?

STEP 4: SET UP FOR TRAINING

There are decisions you must make. One is timing. When? When do you want people to be trained? If they are trained too soon before going live, they may forget the training. If you train immediately before going live, there could be problems if there are training issues. Try to have 1 to 2 weeks between the training and the date of going live with the new process and system.

Where? Where is the training going to take place? Do not use a facility that is remote from where the business process is performed. Try to place the training right next to or even within the area in which the process work is done. This has many benefits. The excitement of the people for the new system will be contagious and carry over to the other staff. Also, people can move right into the process after training. This provides for a closer mental link between training and work performance.

How? How do you want to do the training? You should minimize the formal lecture and concentrate on having them do work hands on. Another suggestion is that they perform the work of the entire process—not just the part that deals with the system.

STEP 5: CONDUCT TRAINING

When conducting training for the first time, make sure that there are team members in the audience who will evaluate your performance. After each half day, schedule some time to review the training for that period. Develop a standardized training evaluation form. The evaluation form should contain their job function, how long they have been in the job, what they found unclear, and where improvement could be made.

Follow up the training by contacting staff after they are working in the new process. Get more suggestions from them as to how to improve the training. This evaluation tends to be much more useful than evaluations conducted at the end of the training.

WHAT CAN GO WRONG?

- People ignore procedures and training until the end of the project. Then there is a rush to throw something together. Quality is not good. The process and system are not easy to use. Get started early.

- The IT staff, not the business staff, are involved in the development of the procedures. During the training, people will start asking for clarification and interpretation. Training slows to a crawl.
- The initial training went fine. However, no one believed about updates to materials or how to train new employees. The result is often that the process gradually begins to deteriorate as the knowledge-level declines.
- The individual preparing the procedures and training materials focuses on the aspects of the system and process that he or she knows well. These are documented in depth. Other areas where there is uncertainty or where testing of the new system failed are ignored. When this is discovered in training, people start to improvise on the spot.

REDUCED SCHEDULE AND COST APPROACH

- For documentation, consider adopting a zero-based budgeting approach. That is, you begin with no documentation and then consider each item in terms of the cost and effort to prepare it and the value in having it.
- Try to get enthusiastic, younger business staff involved in the project. Get them started on doing the combined procedures for the process and system as well as training materials. This will ensure that the materials are consistent with each other and that the documents can be read by business staff.
- Consider the risk and exposure if there is only limited conversion. What if, instead of conversion, you just update the data when the person encounters the database record as part of a transaction? In many cases, this can be an acceptable solution and reduce both the schedule and the cost.

EXAMPLES

For Atlas Bank, the training approach was "train the trainer". Three business staff were identified as having good people skills. They were then trained in the system. After this, they developed training materials. They then trained the IT project team in the process to obtain feedback and validation. The three then trained four other people. Together these seven people trained more than 400 people.

The training technique employed at Atlas Bank was to first give a walkthrough of the process and system. This took about 30 minutes. Then they were trained in groups to process transactions. After 2 hours, they were tested. After testing, they returned to their desks where the PCs had been set up. They entered data to complete the conversion. This accomplished two goals—completed the conversion and reinforced the training.

Millenium Manufacturing believed that they could use most of the data from the current system. Before doing detailed analysis, they contracted with a programmer to write a conversion program. While this was going on, the data analysis revealed the inconsistencies of the information. The conversion program had to change four times. Even then only 30% of the data was salvaged. The remainder had to be entered by hand from paper records.

LESSONS LEARNED

- Try to get several people in the business department responsible for training in the process and the system. Reinforce their importance after the training by stressing the benefits of follow-up training.
- To make the procedures more useful, gather lessons learned on how to do the work and use the system. Add these to the procedures.
- Consider having people who are trained to have a follow-up session after they are working with the new process. During this session, they can indicate what additional information would have been helpful in the training sessions.
- Try to involve people who work in the training process. This will make the training more realistic and valuable.
- After training, make sure that people are immersed in the new process and system as soon as possible.
- Assume the worst when you begin conversion. That is, assume that most of the data in the current system cannot be reused in the new system. This attitude will provide you with a more open attitude toward conversion.

SUMMARY

Many projects in the past have tended to overkill some kinds of documentation and underdo others. In some cases, general documentation policies were applied to all projects and systems regardless of size or scope. The documentation, procedures, and training materials must be adapted to the specific departmental setting. Consider setting alternative levels of effort and budget and then seeing, for each level, what documentation you could generate.

Do not underestimate the value of procedures and training. If there are some gaps in the new system that it cannot handle, then the only structured way to address these is through formal procedures and training in workaround steps. Also, a way to reinforce the killing off of shadow systems and exceptions is to reinforce this in the procedures and training.

WHAT TO DO NEXT

- For an existing system, perform a review of the user procedures and documentation for the business process. Are they up to date and accurate? Do people follow them, or have they developed their own short cuts?
- How are new business department employees trained in business processes? Is the training on the job without any formal instructions?
- Examine a business process for which you can obtain error statistics or in which there is high staff turnover. What is the quality of the procedures, documentation, and training?

Chapter 8

Perform Maintenance, Enhancement, and Production Support

INTRODUCTION

We start with definitions. When a programmer fixes a program to repair an error or increase performance, the work is called maintenance. *Maintenance* in general is work performed to ensure that the system meets the agreed-upon requirements. Examples of maintenance include the following:

- Increasing performance to reduce response time
- Repairing programs that produce erroneous information online or in reports
- Changing an interface in response to changes to an interfacing system
- Making changes for the Year 2000 problem

Many changes are carried out to meet new requirements. Work in this category is called an *enhancement*. Examples of enhancements are as follows:

- Generating new reports
- Extending the system to more business staff
- Adding new data elements to the system
- Implementing new features and capabilities
- Changing a system to accommodate new technology
- Modifying the system to meet new government or union rules

For a computer system in production, *production support* is the effort and work required to perform daily, weekly, monthly, and other annual processing. For an online system, it includes ensuring that the system is available for use. Examples of production support are setting up batch jobs for processing, reviewing the results of production runs, diagnosing problems with the system in production, and taking measures to recover. Performing emergency fixes on a system is considered to be part of production support.

Statistics and studies since the late 1970s indicate that production support, maintenance, and enhancement consume more than one half of the programmers' time. A great deal of the work is unavoidable because a system is affected by many factors outside the control of the system and its supporting staff. Some of these are listed:

- Changes in government regulation (e.g., introduction of the Euro in Europe)
- Union contract changes negotiated during contract talks
- Replacement of current technology with new technology
- Upgrade of system software (e.g., operating system, database management system) to a new version, necessitating changes to the application system
- Modification to an interfacing system requiring changes to the current system

Nevertheless, if no effort is made to control these activities, the sum of maintenance and production support can consume all available time. Enhancements and new development or other work are not even begun. An example occurred in many firms when the Year 2000 problem was addressed. All other work was stopped; new requests were denied.

Maintenance, enhancement, and production support require skilled, experienced programmers or other staff who have in-depth knowledge of the system. As such, their skills and knowledge make it impossible to interchange or move people around based on demand. This means that there is a management issue in resource allocation. For one system, there may not be much maintenance, thus the programmers focus on enhancements. In another area, the programmers are pressed to the wall to deal with production problems. Some business managers perceive that the information technology (IT) area is not well managed and that IT does not assign proper priorities. However, the real problem is the specialization of work due to the in-depth knowledge and expertise required. To understand and better direct these activities, we consider some of the issues that typically must be addressed.

ISSUES IN MAINTENANCE

Most maintenance effort begins with someone in the business department calling IT and indicating that there is a problem. Most firms have the person complete a service request form that explains the problem. After review, the manager, analyst, and/or programmer meets with the business staff to make sure that there is an understanding of the nature, extent, impact, and urgency of the problem. After prioritization, work begins and the system is changed and tested, and the modified code is placed back into production. Sounds very straightforward and sterile. Most want it to be that simple.

- **A business supervisor and programmer negotiate for changes that should be enhancements and make them maintenance to gain a higher priority.**
  Over time, it is natural that a close working relationship emerges between the system supervisors and programmers and the business supervisors and staff. They have worked directly together for years. When there was a simple request, business staff sometimes contacted systems staff directly—bypassing the request process. This is also true of enhancements. In some organizations such as the Roberts Agency, this evolved into a science. Each year the business and systems supervisors would meet and "cut a deal" for work during the next year. This was justified on the basis of expediency, predictability, and the desire to cut through red tape. Although tactically desirable, the net effect was to immobilize system resources and dedicate them away from more strategic work. When management wanted to change priorities, they were told that there could be no response due to these prior commitments.

- **Maintenance changes are not justified. Instead, all requests are ranked and the proper time allocated to them.**
  There is a stack of requests for maintenance changes. The systems manager has several alternatives. One is to spend a great deal of time negotiating priorities with the business department. This then leads to inevitable bad feelings because the people in the department whose requests end up on the bottom of the stack feel alienated, potentially disrupting relations with systems. This occurred in Vision Insurance, where all requests were treated as being equal. The requests were then grouped. The net effect was that there was always a backlog of requests. Frequently, lower-priority requests were handled because they were easier to implement. High-priority requests remained in the backlog queue.

- **Maintenance work is not organized well; thus, the programmer makes many discrete changes to a system that cause problems later.**
  Different organizations exercise different degrees of oversight over programmers. At Atlas Bank, programmers were left alone except for very high-priority work. This situation gave programmers the opportunity to make their own minor changes to the system. In some cases, the programmers made unintentional errors that later led to system crashes and severe production problems. The business department staff got the impression that the system was not stable and could not be trusted. When management jumped in, they focused on repairing the resulting problems. They did not treat the cause of the problem, thus it still continues today.

- **There is no organized approach for performing maintenance. There may be a lack of testing.**
  In many middle and large companies, there is a variety of systems in production. Groups of systems are maintained by different sections of programmers. In some organizations such as Atlas Bank, there was no effort to enforce a uniform approach to performing maintenance. One group had a strong push to exercise configuration management, rigorous testing, and quality assurance. Two other groups had no quality assurance at all. The programmers tested their own changes using the production system. Needless to say, there were cases in which errors appeared immediately after the changed system was put into production.

ISSUES IN ENHANCEMENTS

- **Programmers negotiate for, and business departments agree to, only small changes to a system.**
  Many systems, such as the system at Secour Retailing, were legacy systems that were more than 15 years old. Programmers were loathe to make significant changes due to the risk. When a major change was proposed, the programmers would respond to management that the effort to make the change was too great and the risks too large. Management did not investigate further and backed down. When meetings were held with business departments, both the systems managers and the programmers would explain how complex and risky any changes would be. The business department eventually learned that it was fruitless to propose major changes.

- **Enhancements are not cost justified.**
  A request is made for an enhancement, such as to obtain an additional report. Because of the size of the request (small) or the political power of the business manager, the change is made. The benefits are unknown and there is no cost justification. This happened often in the 1980s when IT organizations reported to the manager of finance and accounting. This manager could almost dictate changes and the IT group had no choice but to comply. Although requests should be justified, there is a better approach that is presented later in this chapter that makes IT more proactive so that political battles for changes can be somewhat mitigated.

- **There is a lack of front-end analysis so that the effort to make the change is far more involved than originally thought.**
  In Millenium Manufacturing, programmers wanted to please managers. When a request was analyzed, estimates of effort were always very optimistic. There was little or no analysis at the start of the work, and the

work always took longer than planned. Instead of insisting on more analysis, one manager just doubled the programmer's estimate—treating the symptom rather than the cause of the problem.

- **Due to time pressure, programmers do not rewrite, redesign, or streamline the program code. Instead, they just keeping adding more code. Old code that was dropped is still resident in the program.**
  The request was approved and work has begun. The programmer is under more pressure due to emergency fixes and other work. The work on the enhancement slips. To catch up, the programmer crams the code into the system. This happened both at the Roberts Agency and at Secour Retailing. In both organizations, programmers never cleaned up the code. The price ultimately paid was high. When new programmers came in to make changes, they had a long learning curve before they could make changes.

- **Requirements are misunderstood.**
  It is the old story carried out in schools around the world. The teacher whispers something to one student who relays it to the next, and so on. By the time the message has gotten around the room, it is garbled and totally different. This occurs where requirements are gathered by systems analysts and then turned over to programmers. The requirements are passed through intermediaries. It gets worse if there is a business department liaison who interprets the request to the analyst who then turns it over to the programmer. There is another problem when there is no feedback where the programmer explains the proposed change back to the business department for validation. The result is not only bad feelings but also extensive rework.

ISSUES IN PRODUCTION SUPPORT

- **Production problems that keep recurring are not fixed. Instead, a band aid is applied each time.**
  Programmers are under tremendous pressure to fix problems with online systems. It could be that there are hundreds of users waiting for the fix. It occurs with batch processing systems where the system has to be run that night. This happened with the payroll system at Vision Insurance. There was no time to systematically research the problem. Instead, a workaround was put into place. After the problem was solved, the programmer was not allowed to go back and reinvestigate the problem. The result is that the problem recurred. Workarounds in the code increased the size and complexity of the programs.

- **Computer operators are allowed to modify the source computer code without the programmer.**
  The programmer works more than 60 miles from the computer center and IT group. It is 11:30 PM and there is a production problem. The computer operator pages and then calls the programmer. Time passes. It is now 1:30 AM and there has been no callback. The operator is desperate and makes several attempted changes. The system appears to work. Unfortunately, the next day it is discovered that the system processed the wrong data. The programmer now has to change the system so that another run could occur during the following night. This happened at four of the example companies. What do you do? Obviously, there is a need for more controls and contingency plans. If there were better documentation, then the operator could take certain steps. Why didn't the companies do this? It seemed routine and not a high priority. After all, the problem was eventually addressed, right? Ignoring the importance of production problems is a commonplace phenomenon.

- **There is a lack of up-to-date production procedures for the network or the system.**
  At the time the network or system went into production, there were up-to-date procedures for operations and for dealing with problems. Over time, the network was changed, updated, and expanded and the system was modified and enhanced. In both cases at Atlas Bank, the procedures were not updated. In response, network and computer operators improvised. Some people on the first and second shifts were very good, whereas people on the third shift made significant errors. There was no effort made to document the approaches and solutions of the people on the first two shifts. The problems continued.

- **There is no requirement for the person who finds the problem to document it completely. It then becomes very difficult to find it later and repair it.**
  In a department in Vision Insurance, an employee's personal computer (PC) freezes up during an insurance renewal transaction. The employee turns the machine off and on. An hour later, the employee reports the problem to the supervisor. The supervisor does not ask about the transaction or the screen. Instead, the system just failed. The problem is reported to IT, where a person is assigned to investigate. The investigation starts several days later. When he or she goes to the supervisor and employee, they do not remember the situation. The search for the problem ends without the problem being solved.

MAINTENANCE AND ENHANCEMENT STEPS

Maintenance and enhancement begin with the request form and proceed through similar steps in terms of analysis, programming, and testing. It makes sense to consider them together. The steps are as follows:

- Step 1: Prepare the request.
- Step 2: Analyze the request.
- Step 3: Set priorities for work.
- Step 4: Implement and test changes to the programs.
- Step 5: Place the modified programs into production and measure results.

For each step, an effort will be made to indicate how to improve the current practices followed in many organizations for the work in the step.

STEP 1: PREPARE THE REQUEST

Almost every organization has a service or system request form. Often, these forms are of limited value because they request business people for technical information or for information that they cannot provide. Here is a usable minimum amount of information:

- Date of request
- Person making request, telephone number, and e-mail address
- Department
- Nature of the problem
- Detailed description of the problem (including screen prints, etc.)
- Impact of the problem on the business if not solved
- Related request (in terms of the business)
- Suggestions as to how to address the problem
- Benefits to the business, if solved

Most of this information is on every request form; however, there are several differences. First, there is the impact on the business if the problem is not solved. This allows the person to state in business terms the affect of not having the enhancement or of not fixing the problem for maintenance. The second item is the benefits to the business if the situation is addressed. It is a good idea to insist on tangible, measurable savings or revenue and not just "ease of use" or some other fuzzy benefit. Together these two fields can help the IT organization assign a priority. Having these two topics on the form also forces the requester to commit to the seriousness of the situation. A form with this information was implemented at Atlas Bank. It was found to reduce the volume of requests by more than 40% because only serious problems or requests were put forward.

After receiving the request, the IT group adds the following data:

- Type of request
- Priority
- Status
- To whom it is assigned for investigation
- Date assigned
- Date that some response is due back to the requester

The type of request can be production, maintenance, enhancement, or other. The priorities can be high, medium, or low. The status can be active, canceled, on hold, or completed.

As work proceeds and is completed, the following information can be added:

- Disposition of request (how the request was finally handled)
- Date request was completed
- Nature of the work performed
- Programs changed
- Databases and files changed
- Verification of benefits

This is much more information than is normally entered. Usually, only the first two items are completed. The third through fifth items are added to support configuration management and the future analysis of requests; they provide traceability for the work. The last item is interesting. In most organizations, there is a lack of accountability of benefits. That is, the benefits are evaluated at the start but never later. This item is intended to enforce some accountability for changes. If a pattern of substantial estimated benefits but few achieved benefits emerges, then this can be employed to scrutinize the estimate of benefits in future requests.

You should consider putting the request form along with online help and assistance on an intranet web page or at least online. All requests should be reviewed by the help desk area and sorted out as to the type of request (production problem, maintenance, enhancement, new development, or other problem).

STEP 2: ANALYZE THE REQUEST

There are two steps to analyzing requests. The first is at the level of the individual request. The second is in the context of ongoing work and the backlog of existing requests.

Analysis of the Individual Request

We assume that the request form includes the information listed in the previous step. The request is typically assigned to a systems analyst. The analyst will review the request and then add the evaluation data elements in the form. Listed here are some specific actions that the analyst should take:

- Go to the programmers and assess what they are currently working on, including the parts of the system they are changing. This can help later if the new request requires changes to the same programs.
- Review the backlog of requests to see if this is a variant of an existing request.
- Review the request in terms of the process plan for the department that submitted the request.
- Visit the business department to examine the request. Here the analyst should verify the nature of the request and add more detail in terms of a potential solution. The impact of the request if not addressed and the benefits should also be verified.

When reviewing the request, the analyst should consider if the request should be expanded or reduced in scope and whether it should be packaged with other changes. This is the reason for the order of the steps and why the analyst should gather data internally first in IT before going to the department. The analyst should also see if the request can be eliminated by changes in procedures. Only as a last resort would the creation of a workaround to the system be considered.

If this were an academic book, there could be a long description of all the work that should go on. In the real world, there is too little time. For the analyst, this investigation is a diversion from bigger projects that have deadlines and due dates. Some consider this an annoyance. How should you try to reduce the time of this investigation? Get to the benefits and impacts as soon as possible. Ask the following questions:

- Why hasn't this change been suggested before now? Changes do not just magically appear. There is often something more fundamental occurring. The request is just a reflection of the underlying change.
- What has changed in the business department that led to the change? Has there been a management shake-up? If it is due to government regulations, then try to get a copy of the regulations to see the actual due date and what compliance is necessary.
- What volume of the total workload of the department is affected by the request? If it only pertains to a small percentage of the work, then even substantial benefits do not warrant doing the work.

An efficient but thorough job is needed here because this analysis is cheap compared with changing things after work has begun.

Analysis of the Requests

Moving up from the level of the individual request, you must now consider the request with ongoing work and the backlog. It may be that there are higher-priority changes waiting in the queue of the backlog. It can also be that the request fits well with several other change requests. You should then create a bundle of requests. Test this bundle idea on the programmers for feasibility. If it is too big, then drop it here. Take the bundle back to the business department for their reaction. There may be additional changes. The order here is important. If you go to the business department first and then to the programmers, you may find that the bundle is not feasible. You then must return to the business department with the bad news. It becomes another case of misunderstanding and raised false expectations.

STEP 3: SET PRIORITIES FOR WORK

Unless absolutely necessary, you do not want to disturb what the programmers are working on. This can be very disruptive because the programmers must shelve all their work without any system change or improvement. At Secour Retailing, this became very frustrating. One programmer had not completed a change in more than 6 months due to changing priorities.

You really want to set priorities for future work. For this you will need to meet with programmers and assure them that they can continue their work and that what is being discussed are future changes. Get the programmers to discuss how they would make the changes in the future. Move toward a bundle of changes so that you can control the application software through releases. A release is the implementation of a group of changes to the system. The release approach has a number of benefits, including the following:

- Improved testing because more effort can be spent in preparing for testing, doing actual testing, and reviewing test results
- Reduced testing than if changes were made individually
- Potentially less risk to the integrity of the source code because the programmer has more time to perform analysis and design of changes

STEP 4: IMPLEMENT AND TEST CHANGES TO THE PROGRAMS

The programmer now makes the changes. The actions that the programmer will take include the following:

- Investigating the source code to see where and how to make the changes
- Designing how to implement the changes

- Making the program changes
- Testing the individual programs (unit testing)
- Making documentation changes in the code as well as in the production operations procedures
- Performing system testing to ensure that the combination of programs performs properly

In a perfect world, you could insist on everything being carried out and full documentation produced. Unfortunately, in the real world this is unrealistic due to both time and work pressures.

What is realistic? In the realm of documentation, the best realistic hope is for comments in the source code and updated operations documentation. Give up on the idea of program and design documentation for the most part. Testing should be done at the system level by an independent quality assurance group within IT. The quality assurance role includes the following tasks:

- Ensuring the integrity of the source code libraries
- Carrying out configuration management for the system
- Maintaining and using test data in test libraries
- Performing integration testing with the test data
- Reviewing test results with programmers prior to integration testing
- Moving the changed programs after testing into production
- Reviewing code documentation and operations procedures

STEP 5: PLACE THE MODIFIED PROGRAMS INTO PRODUCTION AND MEASURE RESULTS

After the changes have been made and the modified system is in production, you are finished, right? Wrong. You must go back and verify the benefits of the changes with the business department. You also monitor the system closely for any errors. It is an unfortunate fact of life that when one change is made in a system, four, five, or more new errors are introduced into the system. This is due to the complexity of the legacy system code, the limited experience of many programmers with the source code (in probably more than 90% of the situations, the programmers did not write the original code).

IMPLEMENT A MAINTENANCE AND ENHANCEMENT POLICY

The preceding discussion provides guidelines for each step of maintenance and enhancement. Overall, it is better to be more proactive and either head off or structure requests for the greatest benefit. After all, the request approach leads to IT being reactive to the request. Frequently, only the "squeaky wheels" generate requests. Others with legitimate needs do not formulate requests. Here is the approach:

- Action 1: Head off requests by holding meetings with business staff to solicit ideas for changes. Allow them free rein to suggest anything. Try to get them to explain the benefits of the change in terms of the impact on the process. Do not push for a specific financial benefit. Also, do not promise anything specific.
- Action 2: Take the list and review it in terms of consistency with the process plan. Then rank these in terms of benefits.
- Action 3: Next, meet with the IT programmers and analysts and review the list in terms of what is necessary to implement and what parts of the system are affected. Throw out the changes that are infeasible.
- Action 4: Group what remains in terms of what exact programs would have to be changed. Now you can rerank the groups of changes based on overall benefit.

There are a number of benefits here. First, it takes a more comprehensive and systematic approach than handling one request at a time. Second, the business department and IT both look at the process and system as a whole as opposed to in pieces. Third, although multiple changes are made to the system, they are in specific programs and have a defined scope. Fourth, you have a better chance of controlling ad hoc requests.

ROUTINE PRODUCTION SUPPORT

A system is in operation. The computer or network operator sees an error message on a console. From the text of the message, it is unclear where the source of the error lies. It could be in the application system. Better call the programmer and have him or her investigate what is occurring. This is just one example of production support.

Other examples of production support come from technology change. Suppose a printer becomes unavailable and the output is routed to a different make and model of printer. The programmer may be called on to insert control code to reroute the printing. In modern systems, this can often be performed by computer operators. For legacy systems, the printing may be controlled deep within the source code. Suppose the data base or files are growing too large. System response time is getting too long. There is a need to compress the files and data for efficiency. The programmer must run utilities to do the compression. Another related example is to archive data to free up space and speed up the system.

Suppose the computer operator makes a mistake and the steps in the system are not executed in order. The programmer must now research the situation to determine the best approach to recover.

With these situations in mind, it is not surprising to understand that routine production support can consume almost one half of the programmer's time. What can you do about this because many situations are not planned and occur through oversight, human error, or other uncontrollable circumstances? Here are some ideas:

- Make sure that computer logs are maintained and reviewed on a regular basis.
- Ensure that the role of the computer and network operations staff is very defined in detail so that they cannot make changes to the code.
- Have the programmer insert controls to protect common computer or network operator mistakes. An example is to give warnings to operators after they enter specific commands.
- Try to make the programmer less available to the operations staff. If the programmer is too readily available, he or she will be contacted as soon as anything occurs. Making the programmer more remote often forces the operations staff to perform more of the diagnostics themselves.

PRODUCTION CRISES

What is a production crisis? It is a situation that causes a system to halt operation. It is also a situation that causes the business department to stop using the system. These crises are unplanned. They can be caused by a variety of things, including the following:

- Changes to interfacing systems that were not communicated to the programmers assigned to the system
- Updates in versions of systems software and utilities
- Installation of new hardware, software, or network products that the system depends on but have now changed

Because these situations are not controllable, what do you do? First, acknowledge that they will happen and concentrate on two areas—prevention, and recovery. For prevention, you must insist that all technology and software changes are discussed and reviewed with programmers prior to their implementation. This will also support testing. Next, it is the job of the IT managers to enforce communications across programming groups. Thus, there must be full disclosure of changes that are being made so that implementation of the change in one system can be delayed until changes are made in interfacing systems and tested.

For recovery, the best approach is to increase awareness that it will occur. Conduct dry runs or simulations of commonly encountered problems with operations staff and programmers present. You can do this in a conference room. Introduce a crisis and then walk through step by step what each person will do. Point out along the way what problems can occur to increase the awareness of miscommunications.

IMPLEMENTING A PRODUCTION STRATEGY

You really want a production strategy to ensure that resources are not bled off on a continuing basis on production problems. Specific steps and countermeasures have been proposed. Now we move up to production overall. Here are some action items related to production that support the strategy:

- Review all production problems with the top 5 systems. Isolate the common, recurring problems.
- Assign teams of two (one operator and one programmer) to see what it would take to make changes so that the problem will not recur.
- Get people back together in a group to go over what was found. Develop an approach to eliminate three problems.

Continue this approach periodically, perhaps twice per year. This should serve to reduce the volume and extent of production problems.

WHAT CAN GO WRONG?

- Different systems are managed differently. There is inconsistency in reviewing and handling requests. More effective and consistent change controls are necessary.
- A critical programmer is about to leave or retire. Yet no one attempts to transition any of the person's knowledge. Often, a junior programmer is assigned the source code and told to go fix a problem. In personnel planning, you should work on the transfer of knowledge from senior to junior staff.
- If you let these areas continue as is, your flexibility and ability to implement major change will be stifled. Note that there will be resistance. People in both IT and business departments will be slow to give up methods with which they are comfortable. To deal with this, point out the benefits that are in their own self-interest. If more major improvements are made, then the business process will improve. If the amount of time in fixes and repairs is reduced, then the programmers can do more creative work. Always emphasize that there is enough work to keep everyone busy.

REDUCED SCHEDULE AND COST APPROACH

The best long-term approach to control changes to a system and ensure that any changes provide tangible benefits is to return to the process plan discussed in Part I. The process plan defined the long-term direction of the business process. What you do is take any requests for enhancements and routine maintenance that are not critical and evaluate these in terms of the process plan. Will the changes provide tangible benefits and advance the process plan? If the answer is no, then either the process plan or the request is at fault. The impact is to reduce the time and cost associated with maintenance and enhancement.

Production support has long been believed to be unassailable because much of the work is "unplanned." However, if you review the logs of problems and operation, you will find recurring problems that likely account for one half or more of the problems. Set up a project to fix these. This effort will then free up time for more beneficial work.

EXAMPLES

The Year 2000 problem and the efforts to fix it has meant that many firms forego any other maintenance and enhancement work other than emergency fixes. It is amazing that the companies survived and prospered. Given that experience, you would think that they would put new controls into place that would not allow such requests to take precedence. But this did not happen. Things went back to the way they were. The same thing has happened in Europe with the conversion to the Euro.

LESSONS LEARNED

- In implementing improvements do not take an ad hoc approach to each situation. Develop a strategy for staging the changes. Measure the results of each change as you go.
- You cannot underestimate the value of measurement. Consider developing a semiannual report on maintenance, enhancements, and production support. Such a report would contain graphs of the total division of labor across IT. Then do the same for each business area. Next, show the change in effort over time. Distribute the report to business managers as well as general managers. What should you hope to achieve? You should achieve an awareness of the effort and resources consumed and the opportunities that are being missed.

SUMMARY

The areas in this chapter are not given sufficient attention by IT management. Yet, they proceed to consume a large percentage of the IT resources. In some companies, the drain is so severe that there is almost no new development and only minor enhancements being performed. If there is one area that a new manager can make an impact on in the short term, it is probably here. If you want to do this, you must take an overall approach. If you only address one area, then the problems will still exist in the other areas.

WHAT TO DO NEXT

- Measure your current maintenance and production support effort. What percent of total resources are consumed by these activities? Compare the percentages by departmental area.
- Review the current system request form. Consider incorporating the additional data elements discussed in this chapter. You will also want to review how requests are handled.
- Given that you want to make some improvements, where do you begin? Select a system that has a business department that is very cooperative and for which management in the department is strong. Begin to implement the steps presented here. You can use this department as an example for others.

Software Packages

Chapter 9

Carry Out Evaluation and Selection

INTRODUCTION

Chapters 9 through 11 address evaluating, selecting, and implementing a major software package. Prior to this, it is assumed from Chapter 3 that the new process into which the software package will fit has been defined. During implementation, there will be references to Chapter 8, which covers conversion, training, and documentation.

In the early 1970s, the world of software packages was different. Most relied on proprietary file structures and not databases. Vendors offered to modify the software at an additional cost to fit your business. Many firms were in the business of providing total hardware and software solutions (i.e., *turnkey solutions*). As time passed, the number of larger software package vendors decreased. The size, functionality, and complexity, as well as price rose. The software packages were rewritten to run under relational database management systems and fourth-generation languages such as Oracle and Informix.

There are currently several definable categories of software packages:

- Utilities, compilers, and other development tools; these are not usually considered packages, but are included here because they can be evaluated as such
- Database management and fourth-generation language software
- Application software
- Specialized software (e.g., mobile communications, GIS = Geographic Information Systems, GPS = Geographic Positioning Systems)

147

- Data communications and network management software (e.g., monitoring, diagnostics, etc.)
- Systems management software (e. g., security, access control, capacity planning, etc.)

The principal focus of this chapter is on application software. However, most of the approach and analysis can be applied to the other categories. Within application software, there are the following:

- Packages that perform a narrow business purpose (e.g., financial forecasting)
- Modular packages in which a company can select any or all the modules from a vendor (e.g., payroll, general ledger, etc.). The modules in such packages are not integrated and you can buy only one if you want.
- Integrated software. The major class here includes Enterprise Resource Planning (ERP) software. These are large integrated software packages such as BAAN, SAP, Peoplesoft, and J. D. Edwards.

Of these, larger firms have tended to purchase the integrated software. Implementing this software is quite risky. Studies show that more than 40% of the projects either fail partially or totally. When there is failure, there is often little to salvage and the firm must write off millions of dollars in wasted effort. In addition, these packages almost require you to hire numerous consultants who are experienced in implementation. In fact, this is so common that it is addressed as a central theme in this part of the book.

SELECTING A WINNING COMBINATION

The purpose of the evaluation and selection of software packages is to find the package that best supports the new business process. However, this is vague. Success will also likely depend on finding the right consultant. Thus, another goal is to find the combination of software package and consulting firm that will implement the software package as quickly and painlessly as possible.

These are two compatible goals. If you select the software package, you might find that there is no suitable consultant who works in your industry segment or that the consultants you are stuck with are not highly capable. Then you are less likely to implement a package that fits the new business process closely.

What would be an example of a winning combination? Here are some ingredients:

- The software package would be in successful use in your industry.
- The software has been through several releases and versions.
- The software would operate on hardware, system software, and network components that are compatible with what you have now.

- The vendor of the software has a presence in your locality.
- The consultant has worked in your industry before with success.
- The consultant has in-depth knowledge of the software package.
- The people that the consultant would assign have experience in the industry as well as with the software package.
- The consultant and the vendor are willing to develop a joint implementation with you as well as a common project management approach.
- Both the vendor and consultant understand what you are trying to accomplish with the new business process and are supportive.

What can happen if you do not get some of these things? Here are some examples:

- The vendor has no staff in the area. Installation and support are much more complex when they are performed through third parties. The third-party staff is not thoroughly trained in the product.
- The software package has been used in your industry but in a different manner or with firms in other regions or under different conditions. The fit of the software is not what you perceived.
- The software package is a good fit, but there are insufficient consultants in your area familiar with the software. You could end up flying consultants in from thousands of miles away.
- The consultants know the package but not the industry. Congratulations! You are about to pay for an extended education.
- The consultant has the experience but likes to do things "their way." You must now spend time trying to reeducate them or attempting to be convinced that they are correct.
- The package as it is delivered is not a good fit. However, the next promised release of the product has the features you want. Unfortunately, it will be 6 to 8 months before you can get your hands on this.

These examples are drawn from real world situations. In most cases, the companies worked hard in the evaluation. What went wrong?

- They did not pick the consultant when they picked the package.
- The firm did not have a new business process in mind. They ended up with a melange of old and new — with few benefits.
- The firm based their software decision on what was promised and not what was available at the time of selection.

SEVEN STEPS IN EVALUATION AND SELECTION

Software evaluation and selection can be viewed as seven steps beginning with finding out what is out there to selecting the vendor and product and entering negotiations (see Chapter 10). The steps are as follows:

- Step 1: Identify software packages. Determine viable software package alternatives.
- Step 2: Identify consultants. Find some of the better consultants for leading software products.
- Step 3: Gather features, transactions, and requirements. Collect information for evaluation and selection support.
- Step 4: Prepare and issue the Request for Proposal (RFP). This is a standard document describing requirements and how vendors should bid their packages.
- Step 5: Evaluate the proposals. Several approaches are examined here.
- Step 6: Make the selection. You will also identify a backup.
- Step 7: Gain management support for implementation.

Notice that the search for vendors and consultants is split into two steps because the method for identification is different. Now suppose that your organization does not issue RFPs. You will still need to give requirements to a vendor for them to evaluate and respond to. The evaluation step (Step 5) is typically divided into parts. In the first part, the field of packages is narrowed to two or three. These are finalists. The second part is an in-depth review of the finalists. There may be a third part that deals with an even finer review of the selected vendor and product.

In the past, it was typical to build lists of features and evaluation criteria and then apply a weighted score method to determine the winner. With this method you apply a score to each criterion for a vendor. Then you apply a weighting factor to the criterion. Sum up the products of weights times scores to arrive at a total weighted score. This method has less value now because many packages have similar features. How do you tell them apart? What additional criteria do you add? You should include having the vendor perform transaction testing or at least explain indepth how their package would handle a range of transactions. This is the approach taken in this chapter.

STEP 1: IDENTIFY SOFTWARE PACKAGES

Action 1: Perform a Software Scan

How do you find what is out there? You can surf the web and get some leads. Obviously, a faster approach is to examine what various firms in your industry are using. You can also get some ideas from magazines that feature articles in your industry. Trade shows are another source of leads.

After you have found a software package, you can collect more information about the vendor and the product from their web site. As you do this, make a note of subsidiary or additional products that relate to the one you are looking for. For example, for one financial software system there were more than 35 add-on software products. Some were PC-based tools; others were host based and extended the capability of the basic package. Write these down and search their web sites as well. When you do make the selection, you will have to estimate the cost and identify what is to be bought. Thus you want to include these too. Some people even set up web sites for comments on the software. Do not neglect these. As you establish files and gather information, consider setting up a computer file for vendor and package information. Sample data elements are given in Figure 9.1.

You should also see what is available regarding software package evaluation using a search engine such as Alta Vista. Some firms offer free evaluation and demo copies of checklists and methods on these sites. There may even be evaluation reports that include the software for which you are looking.

As you are doing this work, establish files for each vendor. Store printouts of web pages and copies of articles. Create a separate file for software evaluation.

Should you contact the vendors directly? There are pros and cons to this action. On the positive side, you will obtain more information. However, much

```
Basic information
    Name of vendor
    Address
    Telephone number
    Fax number
    E-mail address
    Web site
    Name of contact
    Vendor locations
    Affiliates
    Size of company
    Specialties of the firm
    Examples of customers
    Products offered
Product
Name of product
Version/release number
Description
Feature summary
Pricing information
Examples of customers
Sources (allow multiple records)
    Magazine/web site
    Date
    Comments
```

Figure 9.1 Data elements for software packages and vendors.

of this information will just be basic. Many vendors want to visit you and start marketing. Contact means that you come up on the screen of the sales radar. This can then take up much of your time—time that should be spent in project management, gathering requirements, and so on. Hence, try to avoid direct contact with the vendors.

Action 2: Nail Down the Evaluation Approach Details

In this first step, you will also finalize how the evaluation is to be carried out. The steps listed previously give the general picture but not the details. You should answer the following questions:

- Will there be a two-stage or three-stage evaluation? In the three-stage approach, you will evaluate the selected vendor thoroughly before signing a contract. The three-stage approach reduces your risk but extends the elapsed time.
- How will you have vendors show how their software package can handle the new business process? Can they just talk it through or should they write it up in the proposal?
- What will software demonstrations consist of? Do you allow each vendor to demonstrate in their way? Then you will not have a comparison. A better approach is to have them work with the same transactions.
- Who will participate in the evaluation? What will be their roles and responsibilities? Included are business unit staff and managers, information technology (IT), and internal audit.

To settle these questions quickly, you might consider a scenario or possible approach for evaluation. Figure 9.2 contains an example. The roles identify an active role for business staff. Benefits of the new process are identified because you want to determine if the new system will support the new process to the extent that the benefits are achieved. Demonstrations consume time and money to set up and carry out. That is why they are put in the second stage of the evaluation.

If the software you are seeking is a large, integrated package or other complex software, the evaluation will take substantial time and may get politically nasty. It is suggested that you create a formal project plan to manage the effort. Produce regular project status reports. Establish an issues database in which to organize and store problems or potential opportunities that surface. Why make such a big deal out of this? There are several benefits. First, with the plan it will be easier to keep the work on track. Second, you will gain experience and knowledge that should be useful in the implementation. Figure 9.3 contains tasks for a project plan.

Roles and responsibilities

- Project leader/analyst
  Develop project approach.
  Track and resolve issues.
  Review end products.
  Coordinate vendor contacts with purchasing.
- Business manager/staff
  Collect sample transactions.
  Document the new business process.
  Compare the existing and new business processes.
  Identify the benefits of the new process.
  Identify business rules for normal and exception transactions.
  Participate in vendor evaluation.
  Review proposals.
  Evaluate demonstrations and walkthroughs of business processes.
- IT manager/staff
  Document the technical infrastructure for vendors.
  Review software products in terms of fit with infrastructure.
  Determine potential support requirements.
  Review all requirements of the selected system.
- Internal auditor
  Announce and enforce rules for confidentiality.
  Review all documents.
  Sign off on all decisions.
- Purchasing/procurement
  Arrange the mechanics for contacting vendors.
  Coordinate the bidding process.
  Coordinate bidders' conferences.
  Coordinate finalist presentations.
  Support contract negotiation.

Evaluation process

- New process is documented and compared with current process.
- Business rules of new process will be at least partially defined.
- Benefits of the new process are identified.
- Requirements for the new process are defined.
- RFP will contain specific transaction and business rule details, as well as features, evaluation criteria, and so on.
- Vendors will explain how they handle transactions and business rules in proposals.
- Proposal evaluation at the first tier will not include demonstrations.
- Demonstrations will occur for finalists and be geared toward sample transactions.
- Use a three-stage evaluation process.

Figure 9.2 Example of approach for software evaluation and selection.

1000   Project management
  1100   Develop the initial plan
  1200   Identify the initial issues for evaluation
  1330   Implement project reporting
  1340   Evaluation schedule
2000   Roles and responsibilities
  2100   Project leader
  2200   Business management and staff
  2300   IT manager/staff
  2400   Internal audit
  2500   Procurement
3000   Identify software packages
  3100   Identify core software
  3200   Identify additional software—desirable/required
4000   Evaluation framework
  4100   Evaluation approach in general
  4200   Detailed evaluation method
5000   Identify consultants
  5100   Collect consultant information
6000   New process related work
  6100   Identify sample transactions
  6200   New process definition
  6300   Comparison of existing and new process
  6400   Benefits of the new process
  6500   Business rules for normal and exception transactions
7000   Prepare RFP
8000   Issue RFP and receive proposals
  8100   Issue RFP
  8200   Conduct bidders' conference
  8300   Field bidders' questions
  8400   Receive proposals
 9000   Evaluation of proposals—first stage
10000   Evaluation of proposals—second stage
11000   Evaluation of selected vendor

Figure 9.3   Tasks for a project plan for software evaluation.

STEP 2: IDENTIFY CONSULTANTS

Consultants are found through different means than software. If you scan the web for the particular software, you will probably find listings of more than 100 firms. This volume is overwhelming; therefore, you need another approach. The literature both on the web and in hard copy form often contains stories that are example of success. These stories typically identify the consultants who supported the implementation. As with software packages, it is recommended that you not contact the firms yet.

Basic information
  Name of consulting firm
  Address
  Telephone number
  Fax number
  E-mail address
  Web site
  Name of contact
  Firm locations
  Affiliates
  Size of company
  Technical specialties of the firm
  Industry segments specializing in
  Examples of customers
Sources (allow multiple records)
  Magazine/web site
  Date
  Comments

Figure 9.4   Sample data elements of consultants.

Construct a file for each consultant that looks promising. A computer-indexed file is also useful. The data elements are given in Figure 9.4.

STEP 3: GATHER FEATURES, TRANSACTIONS, AND REQUIREMENTS

Here are the actions that support this step. To save time carry these out in parallel.

Action 1: Gather Requirements

This step is critical because it is here that you identify the requirements for the new process defined in Chapter 3. Requirements outside the business process itself include the following:

- Interfaces with other internal systems; for each system, the technical nature of the interface, whether it is batch or online, frequency or volume of the interface, transactions involved in the interface, and error checking
- Interfaces with customers, suppliers, banks, and other firms and agencies; characteristics of the interfaces as above

- Infrastructure
  - Hardware
  - System software
  - Network
  - Database management/fourth-generation language
  - Physical locations and any special facility requirements
- Performance
  - Response time—average time between when the transaction is initiated and when the screen fills in response
  - Throughput—volume of work per unit time
  - Total volume of work
- Size-related characteristics
  - Number of total users (local and remote)
  - Maximum number of simultaneous users (local and remote)
  - Size of databases and records
  - Sizes of data elements (e.g., number of customers, order, suppliers, etc.)

You must validate these requirements. To do this, construct a table as shown here. In Figure 9.5, each requirement is given in the first column. The second column indicates why it is needed. The third is for comments on ranking or need. In questioning a requirement, ask what would happen if you dropped it and the software package you selected failed to meet the requirement.

Do you have all the specific requirements? Here is another suggestion. Construct a flowchart of the new business process. Map each requirement into the steps of the business process. Use a table where the first column contains the steps in the business process and where the columns are individual areas of requirements as defined previously. In the table, you can place an "X" or put in a comment. If there is no requirement, then leave it blank.

Requirement	Why needed	Comments

Figure 9.5   Validation of requirements.

Action 2: Collect Transaction Information

This is a very important action because the transactions you collect may turn out to be the deciding factor in which software package is selected. The word "transaction" can also apply to any piece of work. Listed are some guidelines for selecting transactions and work to be evaluated:

- Select 5 to 10 normal transactions that would represent 70–80% of the workload.
- Identify a transaction for each exception condition. Do as many as possible.
- Describe each workaround piece of work.
- Identify all shadow systems.

Identify these transactions by holding a workshop on the process. After these have been identified, the business department should collect the following:

- Samples of each transaction
- Description of all business rules used to process the transaction
- Input and output for each transaction
- How transactions are edited, rejected, and reworked
- How workload is summarized
- How work is assigned and handed off (this will support workflow routing)
- How transactions are managed and performance is tracked

The business department should also collect samples of forms, reports, input sheets, letters, and other types of input and output to employ later.

All transactions and processing are based on files and databases. For the new business process, you should identify the following:

- All data elements for the new business process
- Edit rules for all data elements
- Characteristics of length, format, and so on
- How data is created
- How data is used, deleted, and updated
- Relationships among data elements

Action 3: Identify Software Features and Contact Vendors and Consultants

Identify Software Features

From the articles and literature you have gathered, you can start to build a list of desired features. Here are some categories:

- General capabilities of the software. This is specific to the software (see Figure 9.6 for a list for accounting software)
- Features and functions
  - Detailed business rules to be supported
  - Government regulations and reporting requirements
  - Union rules
  - Input requirements
  - Output requirements

- Human resources
- Collections on outstanding debts
- Support of the Euro
- Support of multiple currencies
- Support of multiple languages
- Report writing
- Credit management
- Complex cost allocations
- Complex consolidations
- Activity-based costing
- Job costing
- Project costing
- OLAP (on-line analytical processing) support
- Data mining support
- Data warehousing support
- Project management
- Workflow routing support
- Service management
- Sales management
- Time accounting and billing
- Transaction analysis codes
- Treasury management
- Cash management
- Support of work orders
- Support of process manufacturing
- Support of document management and imaging
- Support Electronic Data Interchange
- Support of local government accounting rules
- Support of taxation in multiple countries

Figure 9.6   Examples of functions for evaluating accounting software.

Contact Vendors and Consultants

Be very careful here. You should develop a canned statement that you can use for all vendors. The statement should include the following information:

- Identification of your firm and what they do
- Business process and type of software being addressed
- Evaluation and bidding process to be followed
- What information you would like to initially obtain
- General schedule for the RFP and evaluation
- General schedule for implementation

You also need some information from them:

- Where to send the RFP
- Who the contact should be at the vendor and the telephone number, fax number and mail and e-mail addresses of the contact

The salesperson will likely ask questions to make a first pass to ensure that you are a viable marketing prospect. These people are contacted by many firms and must be selective. Here are some examples of questions they may pose to you:

- Has your company funded or approved money for the system?
- How much money has been allocated? Do not answer this question with a definite answer. Indicate that the company is aware of the cost.
- How will you evaluate the software packages? Indicate that the evaluation will be more than just features.

Make certain that everyone on your team understands that there is to be only management and technical point of contact and one person in procurement. No one else should contact vendors; otherwise, this could lead to confusion and misunderstanding.

Action 4: Define Evaluation Criteria

Software Package Vendor

You can divide the criteria into areas first. Cost has not been included because the cost or price proposal is separate from the technical proposal. Areas are listed as follows:

- *Configuration.* What are your hardware, system software, and network requirements on which the package must operate?
- *Performance.* What are your requirements for response time, throughput (volume of work per unit time handled), error rates, and so on?
- *Capacity.* What is the maximum number of users, customers, suppliers, records, products, and so on that must be supported during a 3- to 5-year period?
- *Support, customization, and modification.* What software support do you require for set up, operation, troubleshooting, maintenance, and enhancements to fit your operating environment? How do you support modification and customization of the software?
- *Requirements for additional software.* What utilities and other software are required to operate the package and get the most out of it? What database management systems or fourth-generation languages are supported?
- *Interfaces.* Identify each interface and its characteristics. How will the vendor supply or support the interfaces?
- *Internal functions.* What capabilities do you require of the software? Figure 9.6 gives a list used by Millenium Manufacturing in evaluating accounting systems for use in multiple countries. Use the results of Action 3 to flesh this out.

Vendor: _____

Criteria	Vendor response	Comments

Figure 9.7   Vendor evaluation form.

- *Data elements for the new business process.* Identify what data are required by the package.
- *Electronic commerce/intranet/extranet support.* What support is there for funds transfer, business-to-business transactions, business-to-customer transactions, report distribution, data entry, and database query?
- *Security and access control.* What security methods and measures are available to support electronic commerce, internal controls, firewalls, and so on?

Prepare a detailed list within each area. Then conduct the appropriate reviews within IT or the business department. After review, transfer the criteria to the first column of a three-column table (Figure 9.7). The second column will be the vendor response and the third column allows for comments or explanation.

Now we turn to the business process. Define a detailed workflow of the business process. Then document each transaction that you want the vendors to test in Action 2. Relate each of these to the detailed workflow. Provide a description of the correct processing of the transaction. Identify any business rules and parameters used in processing the transaction. Remember here that the vendor may say that they cannot do any testing unless they install the software. You will then have them walk through how the transaction will be processed.

Consultants

The evaluation criteria for consultants might include the following:

- General experience of the firm in the industry
- Experience of the firm in the geographic region
- Past projects of a similar nature
- Experience with the software package
- Experience in the industry
- Qualifications of individual consultants
- How the consultants manage their work

STEP 4: PREPARE AND ISSUE THE REQUEST FOR PROPOSAL

Action 1: Prepare the RFP and Bidders' List

Software Package Vendor

Each company has their own "boilerplate" for RFPs or Request for Quotations (RFQs). In general, the RFP should cover the following areas:

- General schedule for evaluation milestones
- Description of the business and business process
- Technology infrastructure
- General functions and requirements of the system (use Figure 9.7)
- Features and data elements of the system (use Figure 9.7)
- Evaluation criteria and method for selecting a winning vendor
- Contact points in the firm
- Outline of the proposal format
- Sample transactions and workflow

The proposal format is important because you seek to obtain standardized formatted proposals to make the review process more efficient. Here is a sample outline:

- General description of the vendor
- General software product description
- Response to general functions and requirements
- Response to specific features and data elements
- Fit of the software to the technology infrastructure
- Technical approach to implementation
- Names, addresses, and telephone number of references for the software
- Software support in installation, upgrading, troubleshooting, and so on
- Pricing of the software
- Sample software license contract
- Sample personnel support contract

Consultants

The RFP for consulting support should contain everything that is in the RFP for the software package. In addition, it should contain a list and description of tasks that the firm expects the consultant to perform as well as what constitutes satisfactory performance.

The outline for the consultant proposal might be along the following lines.

- General description of the consultant
- General company experience

- Experience with the software package
- References to clients in the same industry with the same package
- Technical approach to the work
- Management approach to the work
- Pricing of the services
- Sample personnel support contract
- How the consultant will handle issue escalation, personnel issues, and replacement and substitution of personnel

In the RFP, ask to obtain an original and one more copy than the number of people on the evaluation team.

Action 2: Conduct the Bidders' Conference and Answer Questions

After you submit the two RFPs, you should conduct two separate bidders' conferences. These conferences ensure that people clearly understand what is required and that they know the evaluation method. The meeting format starts with an introduction from the procurement department. Then an IT or business department person steps in to summarize requirements and walk through the RFP almost page by page.

During and after the conferences, questions will be asked. All questions should be written down with their responses. The meeting should be taped so that the answers can be transcribed. Vendors can also submit written questions. The procurement area will then usually issue a list of the questions and answers to the vendors and consultants.

STEP 5: EVALUATE THE PROPOSALS

Action 1: Conduct the First Stage of the Review

Prior to getting the proposals, you should meet with the evaluation team to go over the evaluation process. Create forms for evaluation. On each page, the team members will write down the name of the vendor or consultant at the top of the page and their own name. Then they will rate the package or the consultant against the criteria and make any comments.

Now suppose that you just received copies of the proposals. You open the package and look over the proposal. In some cases, you may have to enter revised numbers to adjust the proposals so that you are comparing apples with apples and not oranges. This is not a brief effort. You will probably have to contact the vendor to ask detailed pricing questions as to what is included, what additional software is needed, and so on. There is a side benefit in that you will learn more about the vendors and their products.

You can now distribute the copies to each member of the evaluation team along with the evaluation checklist. Have the team members send a copy of their evaluations to you so that you can compile a summary evaluation. Make sure that you have each person write down comments as to why they rated a vendor and product the way they did.

Someone should be appointed to contact the references. This should be one person who works off a canned list of questions. During these calls you should cover experience with the vendor and software, how the vendor or consultant performed, what they would do differently if they could do it over again, what lessons learned or hints they might suggest to you, and what surprises were encountered.

As an alternative to telephone contact with references, you can visit them. Assemble a team with an IT member and business staff member. Before the visit, send a list of questions ahead. This will give the firm the opportunity to line up the right people to address the questions. During the visit, you should visit the business departments working with the system as well as the IT department.

Now you are ready for the evaluation meeting. A good approach is to highlight areas of the evaluation where the team members differ. Get a discussion going on each of these. Attempt to gain consensus and then record this. Do not attempt to select a winner. You are only selecting finalists; therefore, you should proceed by process of elimination.

Action 2: Prepare for Final Evaluation

Procurement will now notify the finalists that they have been selected. They will now be given time to prepare a short presentation as well as address how they would handle the transactions, workload, and workflow of the new business process. The presentation for software packages should include the following:

- Very brief history of the firm
- Summary of the proposal
- How the software meets the requirements
- Demonstration of the software with the transactions or walkthrough
- How they propose to attack the implementation both technically and managerially

The consultant presentation might follow this outline:

- Very brief history of the firm
- Experience in the industry and with the software
- Summary of technical and management approach

- Potential problems and risks based on their experience
- How they work with vendors and the firm's IT and business departments

Action 3: Evaluate the Finalists

With the additional input from the presentations and questions asked and answered, the evaluation team can now work through the evaluation. Use the same criteria as before. Identify both a winner and a runner up. You will resort to the runner-up if you cannot come to terms with the winner.

STEP 6: MAKE THE SELECTION

The primary criteria should be the fit with the business process. Second might be the vendor's and the consultant's experience and qualifications. After this, you want to have cost as the third overall criteria. Fourth would be the features list.

After making the selection, it is a good idea to have several meetings to establish a working relationship with the vendor and the consultant. Before a contract is signed, these meetings should be separate. In the meetings, you might want to cover the following:

- How the plan to implement will be defined and managed
- How issues and problems will be resolved
- More detailed questions related to transactions
- Technical qualifications of staff to be assigned
- Realistic feasibility of the schedule

To test the vendor and the consultant, you should also pose some hypothetical but realistic issues and see how they would handle them. Some examples might deal with missing features, lack of fit with the business process, and how to handle disagreements.

STEP 7: GAIN MANAGEMENT SUPPORT

Present informal updates to individual managers in the business department and general management at each step. This will ensure that there are surprises. At the end after selection, you can make a management presentation based on the following outline:

- *Introduction*. What was done. Indicate how much effort was expended. Give credit to the business staff. Summarize the goals of the business process.
- *Winning software package*. Discuss the winning software package and why it was chosen.

- *Winning consultant*. Identify the firm and reasons for selection.
- *Schedule*. Give the future estimated milestones for the contract, installation start, and implementation complete.
- *Cost analysis*. Give a cost analysis of the package and consultant.
- *Issues*. Identify issues that have arisen during the evaluation and how they were resolved.
- *Next steps*. Beyond negotiation, you must come up with a joint implementation plan with the vendor and the consultant.

In all these contacts, you should be positive and neutral in attitude. That is, you are in favor of implementing the new process; however, because you are not in charge of the business department and are not doing the work, you are neutral. People will wonder if you are too positive. It is the business department that should be very positive.

WHAT CAN GO WRONG?

Earlier in the chapter, a series of horror stories were given relating to evaluation. Here we consider what can go wrong in the evaluation and selection process:

- The business staff are not involved sufficiently in the evaluation. They really then are not committed to the software after it is selected.
- The business staff are dazzled by the demonstration of the software by the vendor. It was so slick! This biases the evaluation so that no matter what other information is presented, they still want that package. To head this off, warn them that they might be dazzled.
- People listen to the marketing pitches and see demonstrations. They are inundated with information. They begin to lose sight of the end goal in terms of the new business process. Present the new business process to the vendors at the start so that there is a common point of reference.
- The evaluation focuses on features and capabilities. This becomes dominant in the evaluation. It is a war among products for who has the most checks in boxes. In this situation, it is possible that there will be a tie among products because many have the same features. This can be dangerous because one could be ill suited to the new process.

REDUCED SCHEDULE AND COST APPROACH

To expedite the process, consider doing Steps 1 through 4 in parallel. That is, you would be developing requirements and forming the RFP while you are identifying vendors. A sequential approach can stretch things out from the start. Now go to the last step. Often, getting management approval is a major hurdle because of the cost and schedule of implementation. What can you do to accelerate the review time? Consider giving informal presentations to management and preparing them for the overall cost and schedule as you go.

What can you do to cut costs? See how much the vendor is willing to do to reduce the time allocated to consultants. Another idea is to segment the duties of the consultants to isolate those that require deep knowledge of the software package from those that are just general implementation support. For the general support, you may be able to find other, less expensive consultants. You can also attempt to save too much by cutting the consultants loose too early. Make sure that the package is operational before you release them. Otherwise, it could be expensive to get them back later because they may be reassigned to other projects.

A major method to reduce time and cost is to involve the business staff much more than many projects do in gathering and evaluation of requirements. They will be more appreciative of what it takes to do this. Also, involve them in the trade-off analysis when you are comparing two packages. They must see clearly how a decision was arrived at so that they are behind it.

EXAMPLES

A major European bank purchased an integrated banking software package. Prior to the package, the bank had employed a service bureau operated by another bank. Each time the management or staff asked for some report or change, they were told that the cost and time to implement would be very high. This went on for 7 years. As a result, when analysts tried to define a new process, the business staff had little to offer. They were really out of touch with reality. After a great deal of effort that got them nowhere, the decision was made to select the package based on general criteria and the track record of the vendor to customize (not modify) the software. After selecting the software, it was installed and used to define a new process. Three years later, the business staff were more sophisticated and had a laundry list of new requirements. Management lamented their choice of software; however, it was the right decision given the situation.

Roberts Agency evaluated a number of packages for different functions. For example, they evaluated four packages to replace their personnel system. After doing the evaluation, they found that none matched their needs in terms of policies and procedures. For this area, no software was selected and the current legacy system continued to be maintained. However, they evaluated vehicle scheduling systems and found that one was preferable, but it did not match the process they wanted. Before negotiations, they sat down with the vendor and worked out the new business process as a compromise. The software package and new process were then implemented successfully.

A major aerospace firm replaced its old accounting system with a package after an exhaustive evaluation. The old system could process 3500 accounting transactions in 8 minutes. The new system took 8 hours to do the same work. The package was still not deployed after 18 months. There were major disputes with the vendor. Finally, after 24 months the company decided to scrap the project, sue the software vendor, and select another package. Performance is an important criterion for mission critical applications.

LESSONS LEARNED

- Always assume that it will be a long time before you receive any new software releases. Therefore, you will have to live with the version you bought for an extended period.
- Never alienate any vendor or consultant. You never know who you will need in the future.
- Involve several people from the business department. This will give you backup if some of them leave. Expect attrition given the length of time. You also want to have some people removed if they exhibit too strong a bias.

SUMMARY

Software evaluation used to be simpler. You could select a package based on capabilities and what was available for your hardware platform. The complexity and capabilities of the software have increased, and more is now at stake. There is a higher degree of risk. Management is also aware of the interdependence between the business processes and the software so that the wrong decision could have a substantially negative impact on the performance of the business process. To address these concerns, you should focus on an evaluation method based on business department involvement as well as the new, not the current, business process.

WHAT TO DO NEXT

- Review how software packages are evaluated in your company. Is there a standard approach or does each area have some discretion?
- Consider the results of the software selection process. For several departments review how the people feel about the software after they have

been using it for some time. Do they wish that they had selected something else? Use this effort to collect lessons learned in doing future evaluations.

• Test your ability to do a fast web search. Pick a category of software. Go to your favorite search engine and see what you uncover. Then go to the sites of the software vendors and the data. Also, print lists of related software products. This exercise is useful in that it shows you the extent of what is available and helps sharpen your skills in performing an organized search.

Chapter 10

Conduct Negotiation and Define the Implementation Plan

INTRODUCTION

There is substantial work and coordination involved between the time that the package and vendor are selected and the implementation work begins. The work includes negotiating workable contracts with the vendor and consultant, defining roles for implementation, developing the project plan, and getting agreement on how project management is to be conducted among business departments, information technology (IT), vendors, and consultants. If you do this right, then implementation is easier. If you miss a step or fail to get everyone to agree to an approach, then you will have ongoing, recurring problems. After negotiation and project planning, you begin the implementation.

CONTRACT NEGOTIATIONS FOR THE SOFTWARE PACKAGE

The issues for software packages are very different from that of consultants. For software packages, you are concerned about the performance of the software, whether features are there and work, and similar product-oriented conditions. Here are some critical areas to address as part of the software package contract:

- **What are the hardware, network, and software requirements for the package?**
  Although this was covered in the evaluation and probably during vendor presentations, you want to get it down in black and white so that in case there

169

is a problem, you have recourse. Of particular interest is the detailing of any software licenses required to support the package. There can be "strange" third-party software that only surfaces now. This happened to Millenium Manufacturing. The additional software licenses amounted to more than $15,000 per year. Pay attention to network and software requirements that specify version and release numbers. You may be forced to upgrade to new versions that, if not detected now, can lead to delays and additional expense later.

- **What are the acceptable standards of software performance?**
  Here you want to include the following:

  - Average response time for a specific number of simultaneous users and a transaction set
  - Database and file capacity
  - Throughput in terms of volume of work per second or minute
  - Number of total users that the system can accommodate

  Also, consider conditions related to errors and interfaces, if appropriate. In cases where the package is in widespread use, you can safely reduce this list.

- **What are you going to get?**
  This is the basic software license information. What hardware can you install it on? Can you move the software among platforms without a penalty? How many users are licensed to use the software? How will this be established and monitored?

- **What capabilities is the system supposed to perform?**
  If there are specific functions that you think are important and that the vendor said they support, but are not part of the proposal or written documentation, then you better include these along with a substantial description as to what each means. Experience shows that the truth surfaces here. Some negotiations stop here and the back-up choice is selected because it turns out that you believed or were led to believe that a capability was available now that is really only going to be present in a later version.

- **What tasks and effort will the vendor provide for installation, training, and other support tasks?**
  The vendor should identify the number of hours or days that will be supplied within the cost of the contract without additional expense for the following:

  - Software installation planning, actual installation, and testing to ensure the installation is complete
  - Training of IT staff in the operation of the software
  - Training of end users in the software

- Training for systems and network administrators
- System setup in terms of control tables, security, access control, and so on
- Configuration management
- Other support tasks unique to the package

In addition, go over the terms related to maintenance. How do you notify the vendor of a problem? What is the time allowed for them to respond to start fixing the problem? If the problem is not repaired in a certain number of days or hours, then what is the recourse? What information must you provide to enable them to fix the problem?

- **What documentation will be provided about the system in terms of installation, operation, support, training, and user procedures?**
  Each type of documentation should be indicated along with the media it will be in and any restrictions on copying, printing, and distributing copies within the company. Types of documentation include the following:

  - System installation and troubleshooting documentation
  - System operation manual
  - Network operation manual
  - Software setup procedures
  - End user procedures
  - Training materials

- **What are the cost and availability of staff if additional services are requested?**
  For each category of support, the cost and availability of resources should be stated. Categories include the following:

  - Additional systems training
  - Additional user training
  - Additional setup support
  - Programming services
  - Maintenance support

- **What does the vendor warranty state in terms of the software being free of errors? If errors are discovered, how will the vendor fix these?**
  There are standard warranty clauses for software contracts. Make sure that the steps the vendor will take to resolve errors are identified along with the response time to get on the problem and get it resolved.

- **What are the pricing terms for the various parts of the package? How long are the prices valid?**
  Each package is typically priced in modules. An example might include the basic software, client licenses, additional server licenses, additional software

modules beyond the basic one, and so on. There should be a discount schedule for volume purchases.

- **If, due to the malfunction or failure of the package, the firm suffers financial losses, can they recover damages?**
  Similar to warranty terms, there are standard damage clauses. Leave this to the attorneys. However, you can document the extent of potential damages if severe problems occur.

- **What are the confidentiality terms?**
  There are usually clauses that restrict a company from using or revealing certain information.

- **What happens if the vendor goes bankrupt or out of business?**
  There should be a clause to indicate that the source code of the system will be placed in an escrow account with a financial institution.

- **What are the terms related to new releases and versions? What are assurances that the software package will work with upgrades to hardware, system software, the network, and other software?**
  The vendor probably will issue new releases to the product once per year or even more frequently. What support do they provide for implementing these? Will new documentation, training, and support be made available? If you decide not to upgrade, is there any change in support?
  As the package is used, your hardware may be upgraded as might the operating system, network operating system, database management system, and so on. Does the vendor warrant that the software package will still perform if these are upgrades from current vendors?

CONTRACT NEGOTIATIONS FOR CONSULTANTS

Go to the legal department and obtain their standard consulting agreement. If this is not available, you can ask the consultant for a sample agreement. Here are some of the questions you must resolve in negotiations. Note that there are many other areas that are part of a contract. Space limits the discussion to the most important.

- **What is the consultant to do?**
  This seems obvious, but when you get down to a contract, it is not. You must define the scope of what they are to work on and who they report to in the organization when doing the work. It is recommended that you prepare a list

of tasks to be performed and put these in the contract. There should also be a schedule for these tasks.

- **What are the specific end products that they are to produce? How will these be evaluated?**
  Picking up from the prior point, each task should have an identified milestone. There should be an agreement as to who determines acceptability and what quality is acceptable. It is helpful to include an outline of each deliverable.

- **Who will work on the job?**
  You should push for specific people that you have interviewed in the evaluation being assigned. If you cannot get this, then you should detail the qualifications and background to match the qualifications. You should also have a right of interview and refusal regarding each person on the job.

- **How does a consultant get removed or added to the project?**
  There are several instances where individual consultants must be replaced. One instance is when their performance is not acceptable. A second is when their skills, although suited to early project work, no longer fit the current and future needs. A removal method would be that the project leader from the firm indicates that the person is not to be used in the project after an agreed-upon cut-off time. This time is jointly established because the consulting firm must find a replacement. For replacements, you want to retain the right to interview and try them out prior to approval for work on the project.

- **How is the project to be managed? How is the project plan to be created and updated?**
  The firm has a project leader. The consulting firm must appoint someone who will act as an interface and perform project management tasks for the consulting firm. This is not just a technical question. The overall goal is the same for all parties to work together. Therefore, it makes sense to insist on a central project plan for everyone. With network and internet technology, there are several project management systems that support remote, distributed access and update. Use of this technology should be specified for communications. Frequency of updating the plan as well as modification of the plan should be addressed.

- **How are issues and problems involving consultants to be addressed?**
  Given the elapsed time and number of people involved in the project, there are bound to be problems and issues that arise. That is why you want to define an issue resolution and escalation method for handling problems. The

first level is that of the project leaders. Then the issue escalates up to IT and consulting management. Beyond this is the steering committee.

- **How is additional work approved?**
  The project team or the consultant may identify additional tasks for which the consultant is best suited. Examples might be additional training support beyond what the vendor provides, programming additional functions and reports, and so on. The contract should identify how this work is to be approved. There should be a statement of the work that identifies the tasks and milestones along with a schedule. Then the cost for the work should be identified in terms of labor effort and hourly rate. This information can then be appended to the contract.

- **How will the consultants account for their time and report it?**
  Consultants should complete a weekly time sheet and provide it to the internal project leader. It should give the hours worked each day along with a statement of work and tasks performed. Beyond this, the consultant project leader should provide a biweekly progress report, along with identification and analysis of any outstanding issues.

- **What is the procedure for reviewing and approving invoices? What is the procedure for handling invoice disputes?**
  When an invoice is submitted, the firm and consultant should agree on terms of payment and discounts. Also, there should be a procedure for resolving invoice disputes.

- **What can the consultant reveal or retain in terms of intellectual property?**
  Who has the rights to what the consultant does? What if the consultant invents something in doing the work? Who owns it? What is the consultant allowed to reveal about the work without prior approval of the firm? These questions would be more important where manufacturing or design is being performed. However, they should still be covered here.

- **What is the procedure for termination?**
  If the project is canceled, then can you stop the consultant's work and only for what was performed up to the date of cancellation? Or is there a period of notice such as 10 days? Who can initiate termination? Does there have to be a specific cause?

- **What about liability and damages?**
  These topics are more relevant to software that is a product. However, you should still have these in the contract since the consultant could screw up the implementation by their actions and cause you economic loss.

IMPLEMENTATION PLANNING STEPS

In parallel to the contract negotiations, you should get started on planning for implementation. The steps are listed here:

- Step 1: Identify roles and responsibilities.
- Step 2: Develop the project template.
- Step 3: Identify issues and areas of risk.
- Step 4: Develop the project plan and set the baseline.
- Step 5: Establish change control methods.
- Step 6: Present the plan to the steering committee.

Before examining each step, the overall theme here is that implementation planning is a joint effort with the vendor, consultant, IT, and business departments. You don't want everyone maintaining their own plan. Just reconciling status will drive you nuts and eat precious time. So that everyone is working off of the same basic approach, you should develop a strawman or model for the roles, template, and issues first. This provides everyone with a common starting point.

STEP 1: IDENTIFY ROLES AND RESPONSIBILITIES

Project Leader

The internal project leader does much more than track status and budget and hold meetings. The most successful implementations expand the role of the project leader to include in-depth coordination and problem and issue solving. From experience, you should concentrate on issues. Then the status part will be much easier to do. The project leader should spend more than 50% of their time with the participants and not sitting at his or her desk working on GANTT charts. The coordination and communication role to work with IT, business staff, vendors, and consultants is a major responsibility. It is better to select someone who can work through issues than someone who has just managed smaller projects.

Steering Committee

A large implementation should have a *steering committee*. A steering committee is composed of managers from the business departments affected, IT, and upper management. A steering committee acts to resolve substantive issues that cannot be resolved at lower levels and make major decisions. What should the committee do?

- Review project status.
- Make decisions on policies related to the business process.
- Make decisions on major budget and schedule changes.
- Set priorities for resources.

The people on the steering committee are very busy with their normal work. Keep the meetings to the minimum and have them address issues. Give out information on status separately. The project leader should meet with each of the steering committee members before meetings and between meetings for updates and seek their views on issues. The project leader sets the agenda and timing of the steering committee.

From experience, steering committees can be very beneficial. Some benefits are as follows:

- Provide a means to put pressure on vendors and consultants.
- Provide for pressure to get the business department to change procedures and policies.
- Provide an appeal process for the project manager regarding disputes.
- Give a forum for dissemination of information on the implementation.

However, you can destroy the effectiveness of the steering committee if you involve the people in low-level issues and consume meetings with status. Also, do not bury them in technical detail. When you discuss a technical issue, discuss it in business terms. The project leader can also work behind the scenes with individual managers of the steering committee to resolve issues without involving the full committee.

Not all software packages require a steering committee. You would want to create one if the project schedule exceeds 6 months or some fixed amount of money. You would also want one if there are substantial political issues to be faced or work that involves multiple business departments.

Business Department

In earlier implementations, the business departments had a very limited role. The technology was limited and so the project to implement usually involved IT and the vendor. Today, it is different. The package is going to be the key mechanism to implement the new business process. This implies that there is a need for an expanded departmental role. Some duties are as follows:

- Collect business rules.
- Identify and gather test data.
- Participate in data conversion.
- Perform testing of the software package and interfaces.
- Develop procedures and training materials for the new business process.

- Participate in project management.
- Participate in presentations at meetings of the steering committee.

Each business department should appoint two coordinators. One is a high-level manager to be involved in planning and decision making. The other is a lower-level staff member who will coordinate the performance of the department tasks listed previously. This two-tier approach worked well for Atlas Bank whose senior person also attended the steering committee meetings.

Software Vendor

The complexity and capabilities of newer software has also expanded the vendor role. The old days where the vendor installed the software and then took a walk are gone. The vendor staff must now support the implementation and address interface issues. Vendor knowledge of their software also means that they are included in many package-related decisions. The vendor should identify a project leader to work with the internal project leader. Specific roles for the vendor include the following:

- Project management of their activities
- Planning, coordination, and support of the installation of the software
- Postinstallation testing of the software
- Training of internal technical staff
- Training of business staff
- Other implementation tasks as called for in the contract

Consultants

The consultant firm should appoint a project leader as well. The consultants will be working to support the implementation, make recommendations on issues, and support decision making.

IT Group

The overall project leader is usually someone who knows the business and IT as well. If the project leader comes from the business department, then IT needs to appoint a systems analyst to act as a project leader. The IT group will provide technical support for installation, conversion, interfaces, testing, and other areas.

STEP 2: DEVELOP THE PROJECT TEMPLATE

To get you started and to show what is meant by a *project template*, Figure 10.1 has been included. It is the overall list of tasks for acquiring a software package

Figure 10.1   Sample Software Package Template

ID	O	Task Name	Resource	Start	Finish	Predecess	Resource Names
1		10000 Identify potential software packages and vendors	s Analyst	Tue 10/13/98	Tue 10/13/98		Sys Analyst
2		11000 Data collection on packages from vendor	ys Analyst	Tue 10/13/98	Tue 10/13/98		Sys Analyst
3		12000 Data collection through the Internet	ys Analyst	Tue 10/13/98	Tue 10/13/98		Sys Analyst
4		13000 Data collection through the literature	ys Analyst	Tue 10/13/98	Tue 10/13/98		Sys Analyst
5		14000 Data collection on vendors and consultants	ys Analyst	Tue 10/13/98	Tue 10/13/98		Sys Analyst
6		20000 Evaluation		Wed 10/14/98	Fri 10/23/98	1	
7		21000 Prepare feature and capability lists	t, Bus Staff	Wed 10/14/98	Wed 10/14/98		Sys Analyst, Bus Staff
8		21100 Prepare vendor/consultant checklists	t, Bus Staff	Thu 10/15/98	Thu 10/15/98	7	Sys Analyst, Bus Staff
9		22000 Prepare sample transactions for review	t, Bus Staff	Fri 10/16/98	Fri 10/16/98	8	Sys Analyst, Bus Staff
10		23000 Vendor review/handles the transactions	Vendor	Mon 10/19/98	Mon 10/19/98	9	Vendor
11		24000 Vendor responds on features/capabilities	Vendor	Tue 10/20/98	Tue 10/20/98	10	Vendor
12		25000 Evaluation of packages		Wed 10/21/98	Wed 10/21/98	11	
13		25100 Criteria-maintenance	r, Bus Staff	Wed 10/21/98	Wed 10/21/98		Sys Analyst, Proj Mgr, Bus Mgr, Bus Staff
14		25200 Criteria-benefit	r, Bus Staff	Wed 10/21/98	Wed 10/21/98		Sys Analyst, Proj Mgr, Bus Mgr, Bus Staff
15		25300 Criteria-fit with future process	r, Bus Staff	Wed 10/21/98	Wed 10/21/98		Sys Analyst, Proj Mgr, Bus Mgr, Bus Staff
16		25400 Criteria-technical risk and feasibility	r, Bus Staff	Wed 10/21/98	Wed 10/21/98		Sys Analyst, Proj Mgr, Bus Mgr, Bus Staff
17		25500 Criteria-cost	r, Bus Staff	Wed 10/21/98	Wed 10/21/98		Sys Analyst, Proj Mgr, Bus Mgr, Bus Staff
18		25600 Criteria-interfaces/integration with other software	r, Bus Staff	Wed 10/21/98	Wed 10/21/98		Sys Analyst, Proj Mgr, Bus Mgr, Bus Staff
19		25700 Criteria-consultants/vendors	r, Bus Staff	Wed 10/21/98	Wed 10/21/98		Sys Analyst, Proj Mgr, Bus Mgr, Bus Staff
20		26000 Vendor/consultant implementation support	r, Bus Staff	Thu 10/22/98	Wed 10/22/98	12	Sys Analyst, Proj Mgr, Bus Mgr, Bus Staff
21		26100 Criteria-availability of staff	r, Bus Staff	Thu 10/22/98	Thu 10/22/98		Sys Analyst, Proj Mgr, Bus Mgr, Bus Staff
22		26200 Criteria-quality of staff	r, Bus Staff	Thu 10/22/98	Thu 10/22/98		Sys Analyst, Proj Mgr, Bus Mgr, Bus Staff
23		26300 Criteria-method used by vendor	r, Bus Staff	Thu 10/22/98	Thu 10/22/98		Sys Analyst, Proj Mgr, Bus Mgr, Bus Staff
24		27000 Total evaluation of vendor and package	r, Bus Staff	Fri 10/23/98	Fri 10/23/98	20	Sys Analyst, Proj Mgr, Bus Mgr, Bus Staff
25		30000 Selection and negotiation	yst, Vendor	Mon 10/26/98	Mon 10/26/98	6	Bus Mgr, Sys Analyst, Vendor
26		40000 Project management	yst, Vendor	Tue 10/27/98	Tue 10/27/98	25	Bus Mgr, Sys Analyst, Vendor
27		41000 Project plan development with vendor and consultants	yst, Vendor	Tue 10/27/98	Tue 10/27/98		Bus Mgr, Sys Analyst, Vendor
28		42000 Issues management approach	yst, Vendor	Tue 10/27/98	Tue 10/27/98		Bus Mgr, Sys Analyst, Vendor
29		50000 Initial installation of package		Wed 10/28/98	Wed 10/28/98	26	
30		51000 Table setup	taff, Vendor	Wed 10/28/98	Wed 10/28/98		Sys Analyst, Bus Staff, Vendor
31		52000 Processing of test transactions	taff, Vendor	Wed 10/28/98	Wed 10/28/98		Sys Analyst, Bus Staff, Vendor
32		53000 Analysis of fit with future business process workflow	taff, Vendor	Wed 10/28/98	Wed 10/28/98		Sys Analyst, Bus Staff, Vendor
33		60000 Modifications of workflow		Thu 10/29/98	Thu 10/29/98	29	
34		70000 Modifications/extensions of package		Fri 10/30/98	Fri 10/30/98	33	
35		80000 Data conversion		Mon 11/2/98	Mon 11/2/98	34	
36		90000 System interfaces		Tue 11/3/98	Tue 11/3/98	35	

together with general resources assigned to the tasks. A project template consists of high-level tasks for the entire implementation, a resource pool consisting of general resources and specific people, and the assignment of general resources to the tasks in the template. The template is supported by a project management tool that has been customized for the project.

There are many benefits to using a template. First, it provides a framework for everyone on the project. Second, it helps you to get a head start on planning. Third, it provides a way to establish consistency across multiple implementations. Fourth, it is a useful communications tool with the steering committee.

The project leader should develop the initial cut at the template and review it with each of the other project leaders. This will help pave the way for a common understanding. If the vendor or the consultant resists the template, insist that it be followed. However, they will be able to insert their own tasks within the template.

STEP 3: IDENTIFY ISSUES AND AREAS OF RISK

There are many technical, business, and political issues in a substantial implementation. Some examples from past projects are as follows:

- The department staff will be resistant to the package and to change.
- The consultant staff has limited availability due to other commitments and distance.
- Some of the critical capabilities of the package are not yet in the package. They have been promised soon.
- The interface to a legacy system is a big unknown due to age and complexity.

Once you have identified the issues, set them up in a spreadsheet or in a database. For each issue, find the tasks in the template that correspond to it and put the link in the project management software and the database/spreadsheet. If you find an issue, that is real, but has no task, then expand the template. Now go to the template. Look at each task and see if it has high risk. If it does, see if there is an issue for the task that gives rise to the issue. If not, then create the appropriate issue. This approach allows you to pin down risk and associate risk with issues. It will help you to manage risk better later.

STEP 4: DEVELOP THE PROJECT PLAN AND SET THE BASELINE

With the initial template and issues, you can now ask and assist the department, IT, vendor, and consultant to identify their tasks within the template.

Assign each template task to them so that they can define detailed tasks under these. They can then assign specific resources, define dependencies, and set the duration and dates for the plan. They can also review and refine the issues and tasks that have risk as in the preceding step for the template.

After you have done this, you can set the *baseline plan*. This operation in the project management software copies the start and finish dates into the baseline start and finish fields. You now have the basis for comparing actual progress with plan-to-track progress. Do the same with issues. Include a field in your database or spreadsheet to indicate that the issue was active when the baseline was set. This will support your tracking of issues and their resolution.

STEP 5: ESTABLISH CHANGE CONTROL METHODS

Change control is the method by which you will control scope, additional tasks, schedule, and costs. If you fail to adopt some reasonable approach, you will find that the scope will creep and expand and that control of the implementation will be effectively lost.

A change control method should specify the following:

- How changes can be proposed
- How proposed changes will be reviewed and tracked as issues; here you should define criteria for evaluation
- How the escalation will work in terms of moving up to the steering committee
- How changes will be implemented in the project plan and tracked

You want to allow some flexibility, but you must make proposing changes a lot of work. Otherwise, people will start proposing things that are nice to have but inessential. Make people justify the change and indicate what will happen if the change is not made. You might want to develop a Change Request form. Such a form might contain the following:

- Date
- Person requesting the change
- Title of the change
- Description of the change
- Purpose of the change
- Impact if change is not approved
- Parts of the implementation affected by the change request
- Benefits of making the change

The project leader should work with anyone who wants to make a request and help complete this or try to discourage change.

STEP 6: PRESENT THE PLAN TO THE STEERING COMMITTEE

The baseline project plan, initial issues, and change control approach are presented to the steering committee. Here is an example of the order of presentation:

- Review of the roles and responsibilities
- Summary GANTT chart of the plan
- Detailed project plan for the next 3 months
- Change control method
- Initial list of issues
- Discussion and resolution of a few issues

This last agenda item is very important because it sets the stage and pattern for how the steering committee will function in the future.

WHAT CAN GO WRONG?

- The project leader should be involved in negotiations by doing the footwork for purchasing and legal. Without someone to push along the contracts, staff in these groups will go on to other work. Your contracts could languish for weeks due to lack of coordination.
- The project leader should act as a middleman between the vendor and internal groups and do the same for the consultants. Without someone in the middle, problems tend to escalate to upper management too quickly. This could sour them on the entire project.
- As a project leader, you should develop definitions of roles, templates, and other project management materials on your own and then present them to the people for review and revision. If you attempt to develop the materials with them, it will take too long and could engender bad feelings if there are disagreements. Remember that many people have no experience in package implementation.
- The vendor and/or consultant may resist a common schedule and insist on their own. If you let this happen, you will have problems reconciling the schedules. In the worst case, modify the template to reflect the high-level tasks of the vendor or consultant.
- You do everything right in terms of developing the plan. The problem is that the schedule of the plan goes out too far in the future. It is not acceptable to management and not consistent with what was stated before. What do you do? Analyze the schedule to unearth where the problem is. Work with the appropriate party to simplify the tasks. Get their assumptions regarding the schedule out on the table and see if they can be modified.

REDUCED SCHEDULE AND COST APPROACH

In building a template, contact other project leaders to obtain copies of their project plans. This may save you time in building yours. While you are visiting with them, try to elicit some of their issues and experience. You can then apply these to your project in building the initial list of issues. When you have project meetings to develop the plan, make these working sessions around questions and issues. Avoid status — gather that ahead of time.

To reduce costs, expand the role of the business department since their labor is internal and because an expanded role will lead to more commitment. Negotiate first with the software package vendor to get support for implementation included within the cost of the software package. Stress that they have a stake in the success of the implementation as well as experience. Do not let on that you have a consultant. The vendor may then try to shift the work to the consultant to free up their staff to work on other accounts.

EXAMPLES

Packages were installed by all the example companies. Vision Insurance and Millenium Manufacturing both developed complete plans. Atlas Bank, however, developed a plan without defining issues. Issues started to arise immediately. The project leader had no method for addressing these except to work on them alone. He then started to change the plan based on each issue. Every 2 weeks, the plan changed. Morale in the team sank like a rock. Progress stopped. The vendor assigned people to other clients. There was a crisis that was finally solved by appointing a new project leader and then starting over.

LESSONS LEARNED

- Gather issues from the vendor, department, IT, and consultant at the start — even during negotiations. Start building an issues database right away. This will put you in a better position to deal with the issues.
- As new issues arise, make sure that they relate to tasks. When an issue is addressed, focus on the actions that will be taken and not on the decision. A decision is vague, whereas actions are very specific.
- You will have to work with the department staff during much of your time in planning. You are acting as a mentor and leading them through the method. Remember that you are aiming to get them to participate in detail and not at a distance.

- A consultant may have a canned approach used in other implementations of the same software package. Do not be proud. Review it and see if it fits your organization. If it does, adopt it.
- To keep perspective so that you are not consumed by detail, you should occasionally sit back and ask, "what are the top five risks and issues?" Think about how these could be resolved into critical success factors.
- Try to get members for the steering committee who have some stake and interest in the benefits of the software package. If your organization has political groups and alliances, make sure you leave no group out.
- Plan the interface with the steering committee members carefully. Meet with individual members for updates and to obtain their questions and concerns. Have end users from the business department be part of presentations surrounding specific milestones.

SUMMARY

During these activities to acquire a software package, you have the unique opportunity to build a team composed of internal business staff, IT staff, the vendor, and consultants. After this, you and the team will plunge into implementation. It is your last chance to take an overall view and perspective. You can eliminate many future problems if you create an atmosphere of problem and issue solving.

WHAT TO DO NEXT

- Contact purchasing and obtain copies of software and consulting contracts. Review these and use them to help during negotiations.
- Gather implementation plans for software packages. Also, try to interview people who were involved in implementation and see how they identified and handled issues and areas of project risk.

Chapter 11

Implement the Package

INTRODUCTION

What is your objective in implementing a software package? You might answer with the obvious statement, "Get the software in as fast as possible and in use quickly." This is a true but not complete statement. Doing this does not guarantee benefits that were promised back in early chapters of the book. A better and more comprehensive purpose is to implement the new business process that is integrated with the new software package. You must take this larger, business process point of view if you want the benefits. Keep in mind that the software package itself is a means to an end and a cost. It is not the end. The end is a better business process. It is by no means automatic that if the package is installed, then the process will be changed. In Atlas Bank, Secour Retailing, and Millenium Manufacturing, it was a struggle to then get the process changed. Also, if you do not include process change as part of the implementation of the package, the pressure for the process improvement can die with the completion of the project to implement the software.

The expanded goal has an impact on how the implementation project for the package is structured and undertaken. First, the business department will have to do more work. They cannot just sit on the sidelines and then come in for final testing. Here are some of the activities that business staff can perform with information technology (IT):

- Define the new business process and transactions with the package.
- Identify any shadow systems and exceptions and determine how they will be addressed with the package.

185

- Support data conversion and training as in chapter 7.
- Develop policies and procedures for the new process.
- Participate in defining test cases, testing, and analyzing test results.
- Participate in defining interfaces with other departments and processes.
- Implement the new business process with the software package.
- Measure the benefits of the new process.

If there is an expanded business role, then there must be someone who serves as a coordinator and who is familiar with technology. This could be a systems analyst or project leader in IT. The duties of the coordinator are as follows:

- Work with business staff to get going on their work by generating samples that they can use.
- Review the work of the business staff members and coordinate reviews.
- Act as an intermediary with the software vendor, consultants, and business staff as well as management.
- Develop and maintain the implementation project plan.
- Identify, track, and work to resolve technical and business issues that arise during implementation.

Recall that it was suggested that two coordinators be established in a business department—one higher level and one lower level. The higher-level coordinator can address business procedure and policy issues and questions related to the process. The lower-level person can carry out the actions and decisions reached after resolving the issues.

Implementation of a software package requires decisions on the use of consultants, the degree to which the package has to be customized or modified, and the extent to which the business process can be changed. After discussions of these topics, seven steps of implementation are examined. These are the following:

- Step 1: Install the package.
- Step 2: Learn the software and set up tables and parameters.
- Step 3: Fit the process and the package.
- Step 4: Customize and/or modify the software.
- Step 5: Change the business process.
- Step 6: Implement the new business process with the package.
- Step 7: Measure the results of implementation.

Look at Step 4. There are two verbs there. In this chapter, the word *customize* means to generate reports or change fields within the system. You also might change parameters and table entries. These are okay because they are within the framework of the package. The other is to *modify* the software. Modification used to be common in the early days of software packages when they were limited and the vendor wanted to make you happy. It is very uncommon today for several reasons. First, if you want a change that has real value, it is in the vendor's interest to make the change to the base software and then distribute it in the next release to all users. Second, if the vendor makes changes just for you, then the vendor is only making money on a time-and-materials basis. There is no additional revenue in most cases by selling the changes to others. Third, if there are substantial changes made to the software, you will not be able to accept new releases and versions of the software later. They will be incompatible. Fourth, if the vendor makes changes for everyone, then there will be no baseline product.

MANAGING CONSULTANTS

Large software packages such as Enterprise Resource Planning (ERP) often require implementation support. In Chapters 9 and 10, requirements, evaluation, and selection of consultants are covered. Now they are on board and ready to work. What do you do? From Chapter 10, you have the project plan. Get them started on detailed tasks with milestones that you can review within 2 to 3 weeks. This will give you some feeling of confidence in their work. They will learn how you organize and review their work.

In a large project, issues and problems with consultants are very likely to surface. From experience, here are some common ones:

- The consultant works too closely with the business staff and does not involve IT staff. The problem here is that it is unknown whether the consultant is doing the work for the business staff that should be done, whether the consultant is facilitating the work, and whether there is a transfer of knowledge to the business department. How do you prevent this? Define clear rules on what they are to do with the department. Make sure that IT staff are at the meetings. Have the business staff report on what is going on. Follow up on this in person with them.
- The consultants do not coordinate their work with each other. There is a lack of a project leader on the consultant side. A project leader was there in the development of the plan but has since disappeared. To head this off, establish regular project management meetings with the project leader. Have the project leader participate in defining and updating tasks in the plan. If a problem occurs, then you should meet with the vendor manager and resolve the issue.
- The consulting staff fail to assume accountability for resolving issues. In particular, the consultants neglect to follow up on technical issues that require analysis and contact with the vendor. To prevent this and to resolve it, you should have regular status meetings on issues to apply pressure on them to get results.

THE PACKAGE OR THE PROCESS—WHO WINS?

This is often the basic issue when you implement a package. There is an inherent conflict between the new package and the current business process. On the side of the current process, you have the following:

- Established body of workarounds and shadow systems. Who wants to kill these or change them due to the package?
- Why learn something new? The current process works just fine and has worked for, it seems, a millenium.
- The key of the business lies in the process not in systems, and systems support processes. Therefore, the new package should support the current process.

These are not only compelling intellectual arguments, but they are also deeply held emotional arguments among some users in the business who may feel betrayed by the acquisition of the new packages. Given this attitude what kinds of arguments can they raise? Be ready for these.

- Situations that are odd and very irregular will be presented to you. You will be asked innocently, "How will the package handle this?" What do you say? Probe the frequency of the transaction and how they do it now. Turn the tables on the situation and focus on detail. By working through how the package can handle the transaction indepth, they should gain a better understanding and appreciation of what the package can do.
- If they lose once, they may resurface the same problem in different clothes repeatedly. Follow the same approach.
- They may say that "The new package works fine for normal conditions. For exceptions we still need to use the old procedures." In response, you should have them identify exceptions and work them through the package.

Why do people do this? They may be frightened of change. If you did your job right during the definition of requirements in Chapter 9, this will be less likely to happen. They may be protective of their workarounds and the shadow systems that they created out of their own ingenuity. Give them credit for this and acknowledge that these shadow systems were necessary in the past. Then drill down into the details of what the shadow systems do.

In general, if you do not align yourself to them and make them part of the system, they can sabotage the new package. It has happened before. In one distribution firm, the users docilely accepted the new package. Then they continued on with their current ways of doing business. They put data into the package at the end of the transaction. Many of the benefits of the package were, therefore, never achieved. Are you thinking that this will never happen to you? Forget that. These people have worked with the process for decades. They know what works. Just getting management support is not enough. You must garner support at the bottom.

Who should win? The process should win; however, no one wins per se. You must focus on what will make the business process work in the presence of the package. If necessary, portray the package as a necessary evil during implementation. There is no turning back. You must live with it. Your attitude should be "How can you best live with it?" It will help lead you to trade-offs so that you can get the most out of the software package.

STEPS IN PACKAGE IMPLEMENTATION

STEP 1: INSTALL THE PACKAGE

This sounds simple enough. You have the necessary support infrastructure in place. You have a contract for the software package. Call the vendor and have them install it. However, it is not that simple. Here are some factors to consider and follow up on:

- Installation of the software involves training internal staff in how to do the install and how to monitor and operate the system. Make sure that the internal IT operations staff are trained during the installation.
- There may have to be hardware, network, or other software upgrades prior to installation. Make sure that these are completed and tested before the vendor shows up. When the vendor shows up, verify with them that the configuration is appropriate.
- Installation is never that simple. Your computer and communications environment is unique. Everyone has a unique setting. Therefore, through no one's fault, you are likely to run into problems due to uniqueness. Have your IT staff monitor and work with the vendor on the problems that occur. There will be immediate decisions required. Be present and be part of these decisions. They could affect performance later.

When you install a package, you also typically install utilities to support the package. Give sufficient attention to these so that your staff can use the utilities later. These utilities include security and access control, backup, recovery, and restart. They also include support for batch interfaces with other systems. There may also be a test environment established to allow future testing. The complexity grows if you are dealing with a multiple platform environment where you have to coordinate and link the software.

Define the steps that the vendor will perform prior to their visit. Make sure that internal IT and, if necessary, consultants are on hand to provide support. Plan this down to the hour. Also, define with the vendor how problems will be addressed in advance.

STEP 2: LEARN THE SOFTWARE DURING SET UP OF TABLES AND PARAMETERS

During shortly after the software package is installed, the training starts. There is usually technical support training and end-user training. For ERP software, there is much more involved in terms of training categories and the extent of training. The training is most often carried out by the vendor. Here are some guidelines related to training:

- Select people for training who know the business rules indepth. Make sure of their attendance so that you head off any last-minute no shows.
- Line up people for training in advance in pairs. That is, make sure that there is a backup in case someone leaves or becomes ill.
- Prepare for the training by having the business staff review the current and new business process. Gather some of the transactions used in the evaluation.
- Ensure that business unit staff who will be trained are capable of training others and asking questions that will be posed by other users.
- Have the staff who are to be trained walk through the new business process that will surround the package. This will help to generate questions.
- Outline, with help from the vendor or consultant, specific tasks that the business staff can undertake with the package after training. The most common is that of setting up tables.
- Develop a strategy for how information will be disseminated from the people who are trained to other staff. Make sure that the staff are aware of it.
- Try to send the people to the vendor site or to another away-from-home location. If they are trained at the company or in the same city, there is the likelihood of disruption if they are called or paged about problems in the business department.
- Show an interest in the training by checking with the people halfway through.
- After training, hold some working group sessions where the people who were trained share lessons learned with others. These should also be documented. Both of these steps reinforce the training.

After the initial training and with the package installed, it is time to perform customization (remember, not modification). Most current versions of software packages support customization through parameters and control tables. During the training, each table was explained to you along with the sequential order in which the tables must be addressed. Table and parameter customization are better than doing nothing or changing the source code. However, they are quite restrictive. Examples of table values are as follows:

- Set up and look up tables for the package (part of setup and not customization)
- Selection and customization of reports

- Setting priorities for workload and queueing in the system
- Assigning security and access levels to individual users
- Defining parameters regarding transactions and products supported by the system (e.g., interest rates, etc.)
- Identifying conditions under which letters are generated

There are also tables related to specific business activities such as warehousing, sales, distribution, and so on. In addition, more packages allow you to develop your own letters in word processing and link them in. You can also export output into spreadsheets.

Following are some tips for performing customization:

- Make a list of the tables with the vendor or consultant. Put these in the first column of a table as shown here. In the second column is the purpose of the table. In the third is who is required to set up the table. The fourth column is the status of the table.
- With the tables defined, have people do homework on the table values for each table. Then when they start entering table values, they will be better prepared.
- Make sure that the people learn how to use the tables and make additions and changes.

Table	Purpose	Who is required?	Status

Even with these features, however, packages can come up short. Companies and government agencies that operate in different areas of the world often find that the packages do not fit. Examples are accounting and taxation rules in a specific country. What do you do then? With the packages described here, you would have to hire the vendor or someone else to write custom code to address this situation.

How should you approach this modification and change? Create detailed specifications that include the business rules. Include examples and transactions that can be used for testing. Prioritize these changes based on the timing of when the change is required by the business. After all, some changes may only be required for year-end closing.

Let's consider some tables that specify the functionality and business rules for a specific group of transactions, such as payments. To determine what to do, you must have a business user who is knowledgeable about the business rules. They will have to be removed from their normal job. The person sits with the vendor or consultant and asks the person, "What values do you want?" Panic can set in because there are often multiple answers. If the common situation is considered, then there is one answer. If you consider exceptions, then there is a different answer. Soon the person asks the vendor how the package can address a particular exception situation. The vendor may try to answer without any analysis — a mistake. You or someone must be there to take notes as conversations such as these occur. Issues and questions will arise that are best addressed away from setting up tables.

There is hope for change in the software package industry. The trend toward object-oriented design and development mentioned in the last part of the book is also affecting packages. In the future, you will select a package that will contain libraries of objects such as source code for accounting rules in different countries. If you operate in a country not covered, you will be able to use a "Wizard" and generate code for the additional country. This is not as far off as you may think. It represents a major advance in getting the package and the business process closer together.

STEP 3: FIT THE PROCESS AND THE PACKAGE

As you proceed with the setup and table establishment, you should become more aware of the shortcomings of the software package to address your entire business process. This is true for all packages. It occurs even though you saw in Chapter 9 that the fit with the new business process was of paramount importance. Now that you are involved with the details of implementation, more fine-grained questions will appear.

When these questions and differences arise about the package, remember that there is no turning back. The package has been purchased and installed. It will go live. Now you must dynamically respond to questions and issues about how the transactions can be accommodated in the package. When a question or situation arises, you want to ask the following questions:

- Is the situation that just arose realistic? Under what conditions will it occur?
- How frequently does the situation arise?
- How does the current system and process handle the case now?
- If nothing was done, what would be the impact?

With answers to these questions, you now understand whether the situation is to be taken seriously. For example, if it is a rare event that is not handled by the current system or process, then maybe you decide to do nothing. Here are some alternatives to consider, based on the answers:

- Do nothing. That is, make no effort to accommodate the transaction.
- Attempt to handle it within the package by table values and so on in a standard way.

- Contact the vendor headquarters and talk with technical people there to find out if you can "trick" or deceive the package.
- Try to define a workaround that follows the workflow for the transaction in the current business process.
- Consider using the shadow system that exists (if there is one) or create one as a last desperate measure.
- Contact the vendor for in-depth customization or modification (see Step 4)

There is sometimes a political motivation in people when they suggest strange situations and ask how the package will address it. They may want to keep their current process and system. They do not want to let go. They also may have worked with the system for so long that they can conceive of only one way that the work can be processed. Keep these ideas in mind when you consider this list of alternatives.

As you are constructing the tables and making decisions on the transactions, you are redefining the new business process. You should pause every now and then and review what has been decided so far.

- How far off are you from the original idea for the new process?
- How much of the old process has crept back into the picture?
- What percentage of work has effectively bypassed most of the software package?
- What are some of the complexities that will arise when you attempt to manage the work?
- How will you measure the work if some of it is not performed by the package?

STEP 4: CUSTOMIZE AND/OR MODIFY THE SOFTWARE

After you find transactions that are unsupported adequately by the package, you can consider asking the vendor what they do to customize the software without internal changes. There may be a new release of the product that addresses some of your needs. However, if you are depending on this, do not hold your breath. Neither the time of release of the new version nor the detailed features are locked in cement. The vendor will not be willing to commit to you in writing either.

Consider extensive customization by the vendor to be very undesirable. Ask the following questions:

- Why is your company and process so unique that you require these changes to the software?
- What do other companies in the same industry do who face the same issues? How did they overcome the issue?
- Is there an underlying policy or procedure assumption that is creating this situation?

For certain vertical industries, software vendors are willing to customize and even modify the software. They do this in the belief that most current and potential customers would want these features. Their availability may make the package much more attractive to new business. However, the vendor is placing itself at risk. Each time you modify the basic software configuration, management becomes more difficult. The software becomes unwieldy. Also, the vendor should be reluctant to make major modification that few people would want.

For one insurance company (not Vision Insurance), the vendor agreed to major changes to the basic insurance package. The original package did not work in all 50 states and was too restrictive in terms of insurance agents and brokers. The modified software addressed these issues and provided a much more competitive product for the industry. However, the willingness to make changes does not translate into ease of implementation. The software changes took more than 6 months longer than anticipated. There were threats of lawsuits. Costs rose more than 75% as it became clear that the vendor had underestimated the scope, cost, and duration of the work.

At the end, there is the need to perform a trade-off analysis with all changes that you could dispense of by other simpler means. You are trading off the following with each other:

- Make a system change that may prove very costly and eventually not necessary in the long term.
- Change a policy thereby eliminating the problem and train people in this new policy (simple).
- Modify the workflow in the new process to eliminate the situation.

STEP 5: CHANGE THE BUSINESS PROCESS

With all this work in setting up the new package, you have accumulated transactions that now will be handled differently from the original concept of the new business process way back in Chapter 3. It is tempting to ignore this step and proceed to Step 6 and implement. Do not fall into this trap. You should perform the following actions in this step:

- Review and update all policies that pertain to the process.
- Update all procedures related to individual transactions.
- Define the detailed roles and responsibilities of staff in handling the workload.
- Determine how these modifications will affect how you will measure performance of the process.
- Simulate the new business process with the new system for standard transactions to reinforce the new business process.
- Formally do the training for the new process.

There are several benefits to doing this. First, you have made the process formal. Second, you have identified and highlighted any workarounds and shadow systems. Third, you have pinpointed exactly what people are to do in the process. There is less ambiguity.

A major benefit is that you have paved the way for destroying the old process. Just because you are going to turn off the old system, do not think that that action automatically wipes out the old process. It can linger on in terms of shadow systems. In fact, some people may want to perform both the old and the new process. Without formalizing the new process as was done here, your odds of success at total elimination are reduced.

STEP 6: IMPLEMENT THE NEW BUSINESS PROCESS WITH THE PACKAGE

You have converted the data, established the package, trained the staff, and formalized the new process. You are now ready to go live. How do you do it? You have several alternatives:

- *Go live with a specific division or organization.* The argument for this is that you can take experiences in one location and apply it to others. The counterargument is that this stretches out the implementation and makes the process harder to manage with two processes at the same. This alternative is the preferred one when you are deploying a new process to hundreds of retail stores. For example, it was used at Secour Retailing for point of sale, bar coding, and scanning.
- *Go live with specific sets of transactions.* This is a good approach if you are waiting for the vendor changes to be ready to handle exceptions. You could do the exceptions later. The negative is that you are operating two processes. This may incur more errors. This was done at Millenium Manufacturing where exceptions were moved into part of the production line away from the main workflow.

- *Go into production with the entire process for all transactions.* This is generally preferred, but it may not be possible if there are many locations to be implemented. Total cut-over was carried out at Vision Insurance.
- *Have only a part of the process go live (e.g., data entry).* This is the only way to proceed if you have a process in which data are captured at the early stages of the process and then queued up for review later in the process.

Obviously, your decision will depend on the situation. Note that doing the old and new process in parallel is not listed as an option because resources are insufficient to do this.

STEP 7: MEASURE THE RESULTS OF IMPLEMENTATION

The measurement of benefits of implementing a new system and process have been addressed and are considered in Part IV. Here we concentrate on some of the other measurements:

- Vendor and package performance
  - What features were promised in the software but not delivered?
  - What are the missing features and capabilities?
  - How responsive was the vendor to questions and problems?
  - What was the quality of training?
  - What was the quality of documentation?
  - How well did the vendor work with the internal staff and consultants?
  - What are the unresolved issues as of today?
  - How did the vendor manage their own people?
  - What was the level of the vendor technical support?
- Consultant performance
  - What was the technical quality of their work?
  - Did they employ the same staff throughout the project?
  - How responsive were they in responding to issues?
  - How well did they get along with the internal staff and vendor?
  - Did they transfer knowledge to the internal staff?
  - How did they manage their own people?
  - What was their level of technical knowledge about the software product?
- Lessons learned
  - What turned out differently than anticipated?
  - If you were to do it again, what would you do differently?
  - Was there an effort to apply lessons learned as the project progressed?

- Business department performance
  - How involved and committed were the staff to the package?
  - Was the department reasonably flexible with respect to process changes?
  - What was the level of staff participation?
  - What was the level of management commitment?
  - Can the business department address table changes?
  - What was the quality of procedures and policies? Are these clear?
  - What was the quality of training?
  - Is there an effort to capture and answer *frequently asked questions (FAQs)*?
- IT performance
  - What was the quality of IT support?
  - Do the IT staff know how to operate the software package?
  - What is the quality of the interfaces between the package and other systems?
  - What errors have occurred since going live?

Why do this if the project is over? Because you want to work for cumulative improvement. You also want to make any adjustments in roles and procedures. You can carry out the measurement by converting the items listed here into a checklist and then rating them on a scale of 1 to 5 (1–low, 5–high). Next, you can validate what you have done by reviewing the scores versus your overall memory and experience.

IMPLEMENTING NEW SOFTWARE PACKAGE RELEASES

The vendor announces that a new release of the software will be available in 3 months. Will you install it? This is a joint business and IT decision. Some firms are very conservative and always stay one release back. The reason is that they will not incur the problems and errors of something new. The problem with that approach is that they are missing potentially major new features that would help the business.

How do you make the decision to go with the new release? There are two aspects to this decision:

- *Technical.* Does the new release offer capabilities that improve performance and operation support? Does the release support other software that you have?
- *Business.* How do the new features of the release translate into the business process effects? What are benefits of the features and to what percent of the work do they apply?

It sounds like a simple trade-off of what do you get versus what does it cost in effort. However, there is a factor to encourage you to move to the new release—vendor support. The more you lag in releases, the less likely the vendor is to provide support. They may believe that you are not keeping up your part of the bargain by not upgrading. This will come through in italics if you complain about software problems that are fixed in the new software release.

At Atlas Bank, they decided to always stay one release behind the current version. This ensured that they were not pioneers. This approach also delayed them from using the enhancements of the new version.

IMPLEMENTING ERP SOFTWARE

ERP software is one of the most complex to implement. In some of the packages, there are more than 1500 interrelated tables. Implementation can take several years due to complexity and the scale of work required. ERP software consumes both consulting and internal resources.

Some of the packages are such that you must fit your processes to conform to the basic software, even with tables. This works best if you have business processes that people view internally as terrible. Then the software package provides a new process through its structure. Some companies embrace this idea when they have widespread divisions in various countries with different styles of doing business. Having standardized ERP software in place also eases the pain when companies merge.

What are some guidelines in implementing ERP? There are probably hundreds, but here are three that seem to recur:

- Plan the use of senior business staff very carefully so that they can be shared between the ERP and the current process.
- Be willing to make major process changes to accommodate the package.
- Concentrate on implementing one module throughout the firm, if possible, instead of implementing all modules in one location or division.

The ERP requires massive involvement of internal staff. Consultant support can help but even then the demands are great. Do not allocate someone 100% of the time to the ERP. They will become too isolated from their old department and lose touch with the processes.

Furthermore, you have to be flexible. There is no software modification allowed. ERP software is totally encompassing almost all work. Thus, it is much more difficult to create workarounds and shadow systems.

As an example of the third point, an office products firm implemented a major ERP package in one division. It was a total failure because the implementation could not be spread to other divisions. Moreover, the high cost was a huge burden. After spending 18 months and more than $15 million, the project was abandoned. Nothing was salvaged.

WHAT CAN GO WRONG?

- Some business staff will not give up on the notion that the old process and system were better than the new package and system. Be on the lookout for people to try in small ways to keep the old process even if they cannot keep the old system.
- The vendor and the consultant may get into conflicts and disagreements. You should expect this to happen because they arrive at the same place from different directions. The vendor tends to defend the package. The consultant criticizes it because he or she has to implement workarounds to handle the shortcomings of the package.
- One factor that seems to doom package implementation is elapsed time. If the implementation drags out, then the business staff must return to the business process. The IT people are redeployed. The vendor moves on to other customers. The consultant money runs out. Why does this happen? There can be overriding issues that the project team cannot solve and that management cannot address. There can be a fundamental mismatch between the process and package. People believe that the safest thing to do is to let it die a slow, quiet death.

REDUCED SCHEDULE AND COST APPROACH

There are a limited number of ways you can try to reduce the time and money of implementation. Here are some guidelines:

- Try to be very strict that no changes in the package will be tolerated. This will cut down on customization.
- Aim for pushing transactions through the package as soon as possible. This accomplishes several goals. First, people get more hands-on use — reinforcing training. Second, you will address problems with transactions earlier.
- Be willing to change policies and be flexible. Policies are cheap to change as long as the changes are valid.
- If you have multiple locations, then you should involve business staff from all major locations in the initial implementation. This will reduce the installation time when the other locations get the system later.

EXAMPLES

All the example organizations implemented software packages. The critical success factor in all cases was whether the business department worked to mesh the process with the software package while still keeping the real world in mind. We review what happened in several cases. Millenium Manufacturing implemented an Manufacturing Resource Planning (MRP) package. Everyone welcomed it when compared with the old system that it replaced. They tried so hard to use it as is, that they force-fit all work into it. After several months, people realized that some changes were needed. These changes were made through customization, not modification, of the software.

The Roberts Agency acquired software for timekeeping in the 1970s. This was before most of the current technology was available. There were no PCs installed. The programmers were very creative and developed new applications inside the package. They added an electronic broadcasting capability for getting messages to bus drivers and added a bus driver sign-on system. The internal code was modified for evaluating and then implementing new union contracts. Within 3 years, Roberts was unable to accept any later releases from the vendor because the software had been severely modified. One of the programmers quit after 5 years and maintenance was reduced. After this, when the business department or management wanted changes made, they were difficult, if not impossible, to implement. This situation continued for years because there was no software package similar to the system.

At Vision Insurance, the head of accounting went out and purchased a cheap off-the-shelf general ledger package. He then contacted IT to install it. The package required network components that Vision had never used. The cost of installing two PCs, the server, and the network for this "cheap" package was more than $30,000. After all of this effort, the package was never fully implemented. The head of accounting lost interest and people returned to their old ways. Eventually, the little network degenerated into word processing.

LESSONS LEARNED

- The basic software package can be the tip of the iceberg. There may not only be additional modules for you to buy from this vendor, but also products to purchase from other vendors to get additional functions.
- When you are getting familiar with a new software package, it is more important to understand what the package cannot do than what it can do. If you focus on what it can do, then when an issue in the business process arises, you are likely to try to shove it into the package without analysis.

However, if you find out something it cannot do, then you can immediately get people thinking about how to deal with this. You raise the level of awareness of the package's limitations.

- In the training from the vendor, try to get hands-on training with the package. Make sure that people in the audience are relating what they are learning to the business process. Often, there is so much pressure that people just concentrate on how to work the mouse or keyboard and are not thinking of the business process implications of what they are doing.

SUMMARY

Most packages do get implemented. However, there are degrees of success and failure. You can implement the package and still leave the old process. Benefits, as a result, are greatly reduced. At the other extreme, you can force everything through the package and not allow for any customization. Then people will start creating workarounds and shadow systems again. To avoid these extremes you must be constantly in a trade-off frame of mind. That is, you must be flexible and willing to consider anything if there are sufficient benefits because your overall purpose is to make the implementation a success.

WHAT TO DO NEXT

- Within your organization, visit two departments that are using packaged software. See if you can detect workarounds and shadow systems. In each department, try to piece together how the process and the package were fitted together.
- Try to capture lessons learned from software package installations in the recent past. You will find that many of the ideas still apply.
- Use the measurement approach in this chapter to measure a process that employs a software package.

Part IV

Knowledge Management

Chapter 12

Identify Opportunities and Develop Your Strategy

INTRODUCTION

DEFINITIONS

What is knowledge management? *Knowledge management* consists of systems, information, and processes that take information and turn it into structured knowledge to support specific and general business purposes. To better understand this term, let's consider some examples:

- Most major grocery store chains build large databases out of their sales data. The sales data include what people bought, where they purchased it, when they bought it, and what they bought it with. If the customer used a "club" type card, then they know the age, sex, and some demographics of that person. This gold mine of information is analyzed internally by the store to see what things should be placed with each other and how to market, develop marketing strategies, and so on.
- The information from the grocery stores is sold or traded to manufacturers that supply the stores with the goods. They analyze the data to determine advertising strategies, advertising effectiveness, which brands appeal to whom, and so on.
- The U.S. Defense Department and other government agencies promote lessons learned. That is, they promote the gathering, organizing, and dissemination of knowledge on how to perform specific activities. For example, there is an entire database on how to repair and cope with problems related to army trucks.

- Banks and other financial institutions have implemented behavioral scoring for processing credit applications. Behavioral scoring takes as input a large number of items from your credit card file and then analyzes this along with the credit application information against a huge information store of credit scoring methods. Another banking application is the detection of potential credit card fraud. Knowledge management applications can cut fraud losses.

There must be a caveat here. This is a management book and thus there is no indepth technical discussion about specific technologies or about the technical details involved in establishing the database.

These are just a few examples of how organizations are employing computer systems in knowledge management. Knowledge management applications are different from other types of systems, such as traditional database management, in several ways:

- In a data base application, the information is intended for use by a specific application. In knowledge management, information or knowledge is intended to be employed for a variety of purposes—many of which are not defined when the data are being collected.
- In a standard online or batch application, there is structure emanating from the specifications and design. In knowledge management, there are tools used to extract, analyze, and synthesize the information.

Table I provides a comparison of traditional computer applications with those of knowledge management.

Why is knowledge management important? This is one of the fastest growing areas of information technology (IT). Hardware advances make larger databases and more data analysis possible. With every new technology such as electronic

Table I

Comparison of Traditional Systems with Knowledge Management

Factor	Traditional	Knowledge management
Volume of transactions	High	Low
Number of users	High	Low
Sophistication of users	Low	High
Purpose	Efficiency of processing	Effectiveness in data analysis
Access to system	Structured through menus and transactions	Unstructured, ad hoc
Tools	All business rules are preprogrammed	Wide range of statistical and analytical tools
Amount of data accessed	Low for one transaction	Very large
Output to end user	Specific for transaction	Variety of reports and summary information

commerce, there is a knowledge management component. In electronic commerce transactions, you know a great deal about customers — information and knowledge that you can use later for sales. Many large and many medium-size companies now have knowledge management groups. You can also ask the reverse question. What would companies do without knowledge management? They could make intelligent decisions about new products and models. They would have to revert to much more labor-intensive methods.

A *data warehouse* is the central store of the information and knowledge. Data warehouses do the following:

- Enhance the presentation and analysis of information by having the information in a single location
- Support analysis of data to determine facts and truth about the information
- Support the analysis of vast amounts of data for underlying relationships

A data warehouse draws from data in diverse files and databases and makes much more analysis possible than with a single application system. Otherwise, you would have to write interfaces for each situation you wanted to consider. A basic tool for analyzing relationships within the data warehouse is *on-line analytical processing* (OLAP) software. Suppose you think that people who buy a certain kind of laundry detergent also buy a specific brand of bleach. Using OLAP you could determine the truth of this hypothesis.

When you think of a data warehouse, you are looking at potentially hundreds of gigabytes of information. Think of a table with millions of rows and columns. How can you analyze all these data? This is the area of another tool of knowledge management—*data mining*. In data mining, you analyze the data to determine underlying relationships. Unlike OLAP you do not need a specific idea to test at the start. Data mining uncovers the relationships by itself through extensive statistical and numerical analysis.

A data warehouse serves as the basis for many applications to aid in decision making and understanding. *Decision Support Systems* (DSS) aims to support decisions in specific areas such as loan approval. *Executive Information Systems* (EIS) use the data warehouse to generate reports based on a variety of questions and requests.

BENEFITS AND PITFALLS

There are a variety of benefits of data warehouses and related tools:

- There is support for consistent analysis of business processes across an entire organization.
- Integrated information across the company can replace many department- or division-level efforts.

- Analysis of many complex business decisions is made possible.
- Data warehousing is an approach that can cope with the explosion of information within the organization.
- Data warehousing becomes an essential business tool to remain competitive if your competition is using the technology.

A recent study indicated that the return on investment of data warehouses can exceed 500%.

With all of these benefits, why do many data warehousing implementation efforts fail? Listed are some common reasons:

- Companies tend to underestimate the cost and effort to build the warehouse. They run out of money before the warehouse is completed.
- The existing files and databases are in such poor shape and incompatible that a major cleanup is necessary. Management loses patience.
- The scope of the data warehouse grows as more people see its benefit. The expanded scope totally overwhelms the project.
- The payoff for a data warehouse only occurs when you have collected the information and established the infrastructure. Again, management loses patience as they seek shorter-term results.
- No thought is given to what will be done after the data warehouse is set up. The objectives appear fuzzy. Management has not supported setting up the infrastructure for analysis.

If the existing systems are not compatible and you do not have a large budget to build the data warehouse, what can you do? You could collect data from several application systems and establish a large database for analysis (VLDB, Very Large Data Base). When you do this, the resulting data base is called a *data mart*. A data mart is not as comprehensive as a data warehouse, but it can provide useful results at a more affordable price in less time.

You have been introduced to some of the concepts related to knowledge management. Here are some more that you will encounter:

- *Drilling down* through the data. This refers to the fact that when you begin to perform analysis in a data warehouse, you work from a high level of information to more detail.
- A data warehouse depends on knowing all about the data elements in the data warehouse — their characteristics and where they came from. This information is called *metadata*.

ISSUES IN KNOWLEDGE MANAGEMENT

Before you plunge into selecting potential applications, making a decision on a data warehouse and so on, you must face the following issues:

- What is the condition of the information that you have now?
- What are some tactical questions and applications of a data mart or data warehouse that you want to address in the near future? This will provide a focus for your effort.
- Who owns the information? Who will own and operate the data warehouse? These are not idle questions. Ownership must be at one of the highest levels of the organization.
- How will the data warehouse be created and updated? How will updates be coordinated and controlled? Answering these questions means that you will implement controls on operational systems that feed the data warehouse.
- What tools and approach will be used to create the warehouse, analyze the data, and perform maintenance of the data warehouse?
- Who will have access to the data and tools? Answering this question addresses a fundamental issue of control.
- How will shortcomings of the information be addressed?
- Business managers have a need to analyze historical data. The systems that will generate the data for the warehouse have been modified over time so that the historical data for a system is often not compatible with the database structure of the application system today.

STEPS IN ESTABLISHING KNOWLEDGE MANAGEMENT

Five steps to establishing a data warehouse are identified as follows:

- Step 1: Identify knowledge management opportunities.
- Step 2: Assess the current state of the systems and data.
- Step 3: Determine your knowledge management strategy.
- Step 4: Select the appropriate methods and tools.
- Step 5: Implement the knowledge management database and the initial application.
- Step 6: Fine tune your approach to knowledge management and expand the applications.

This chapter focuses on the first three steps. Chapter 13 addresses Step 4, whereas Steps 5 and 6 are covered in Chapter 14.

STEP 1: IDENTIFY KNOWLEDGE MANAGEMENT OPPORTUNITIES

To be successful, you should approach your goal from different perspectives:

- *Application software perspective.* Look over your current applications. Which ones might have useful data that could be extracted and transformed into a database without massive programming? The most likely candidates tend to be those newer systems that are based on database management systems or fourth-generation languages.

- *Business department perspective.* Review an organization chart to determine possible candidates for knowledge management. Do not restrict your attention to financial, planning, and accounting areas. There are some good opportunities in operations such as warehousing, manufacturing, sales, and distribution.
- *Business process perspective.* Review the critical business processes and examine how each is measured in terms of performance. Also, consider what summary and analytical information the systems for these processes provide.
- *Management perspective.* Consider what types of questions upper-level managers might ask about operations in the firm. These questions can then be employed to define opportunities.

From experience, some of the best opportunities occur in the business process area. These also tend to yield regular benefits once implemented.

An opportunity for what? Knowledge management is a very general area. To identify an opportunity, you must identify the following:

- Source of the information that will feed the knowledge database
- Users of the knowledge database
- Tools required to analyze the information
- Complexity of establishing interfaces between existing systems and the new knowledge database
- Level of skills required to use the tools and the level of current staff
- Potential frequency of database use
- Rating of potential data quality and completeness of the information
- Estimate of potential benefit

Now create a table. The rows are the bulleted items listed previously. The columns are the potential opportunities. In the table, enter 1 through 5, with 1–poor and 5–good. For example, if the source of information is easy to get to and has valid data, items 1, 4, and 7 would be rated high. If the systems that would provide the data are legacy systems that are very complex, then items 1, 4, and 7 would be rated low. An example for the Roberts Agency is shown in Table II. Rather than numbers, comments have been inserted in this table.

Tables and charts are useful here because imagining a huge data warehouse with many dimensions is almost impossible. Next, create another table in which the rows are the potential opportunities. The columns represent examples of questions that could be addressed by the knowledge management application. Additional columns would represent the benefits of answering the questions.

Table II

Opportunity Evaluation for the Roberts Agency

Criteria	Lessons learned data base for bus drivers	Data mart for bus maintenance	Data warehouse of financial infomation
Source of information that will feed the knowledge database			
Users of the knowledge database			
Tools required to analyze the information			
Complexity of establishing interfaces between existing systems and the new knowledge database			
Level of skills required to use the tools and the level of current staff			
Potential frequency of use of the database			
Rating of potential data quality and completeness of the information			
Estimate of potential benefit			

This table will support your rating of potential benefits in the first table. The table for the Roberts Agency is given in Table III for two questions.

It is useful to create a chart showing some of the dimensions of the analysis. Figure 12.1 is a spider or radar chart for the factors in the bulleted list for the Roberts Agency. This proved very useful in evaluating alternatives later.

Table III

Opportunities and Questions for the Roberts Agency

Opportunities	Question 1	Benefits 1	Question 2	Benefits 2
Lessons learned data-base	What if the fuel pump breaks?	Reduced damage	What are the most dangerous intersections?	Safety
Data mart for bus maintenance	What is the most expensive type of bus to repair?	Better maintenance planning	What is the trend of prices from specific vendors?	Vendor analysis
Data warehouse	What is the trend of department expenses?	Better understanding of budget	What is the effect of improving benefits?	Better estimation

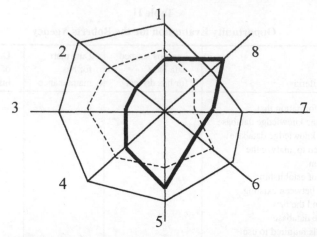

Figure 12.1   Spider chart of opportunities 1—Source of the information that will feed the knowledge database, 2—Users of the knowledge database, 3—Tools required to analyze the information, 4—Complexity of establishing interfaces between existing systems and the new knowledge database, 5—Level of skills required to use the tools and the level of current staff, 6—Potential frequency of use of the database, 7—Rating of potential data quality and completeness of the information, and 8—Estimate of potential benefit. The solid line is for the lessons learned database. It requires few tools, no interfaces, and little skill, and has a substantial level of potential use and some benefit. The dashed line is for the data mart, and the heavy solid line is for the data warehouse. The data warehouse has more benefits, simpler interfaces, better quality data, and more experienced users than maintenance.

How do you verify what you have done? Well, you cannot go out and conduct a pilot or prototype. However, you can carry out a simulation of the steps using the knowledge management application. This allows you to simulate the benefits. It will also aid you later in determining what tools you will require.

What do you present to management? You have the chart and tables. Here is an example order of presentation:

- Introduction—how data warehousing and knowledge management are used in the industry
- List of potential opportunities with substantial possible benefits
- List of criteria (bulleted list)
- Tables and chart
- Recommended actions

The main recommendation is that you pursue the next step in the analysis. You are not asking for approval of a project. It is too early for that.

STEP 2: ASSESS THE CURRENT STATE OF THE SYSTEMS AND DATA

Before you jump in and go for some opportunity, you should review the current state of the technology and data that are in your systems. There is a direct benefit to this even though you may decide not to implement knowledge management. The work may lead to a recognition of the shortcomings of some of the systems and their impact and restrictions on the organization and management. In the case of Vision Insurance, such an analysis revealed that the claims system badly needed replacement. It was very difficult, if not impossible, to employ the claims data in setting rates for insurance policies.

How do you undertake such an assessment? Start with a list of criteria for a single system:

- Age of system
- IT opinion on the condition of the system in terms of being able to make changes (based on maintenance and enhancement experience)
- Completeness of the data in the files and databases (Are there data elements that people need but that are not present?)
- Adequacy of the data elements in the system (Are data element lengths and characteristics too limited? Examples are truncated fields, 5-digit ZIP codes, telephone numbers without area codes, etc.)
- Accuracy of the information
- Effort to establish interfaces (What effort did it take to do this last time?)
- Timeliness of the information (How frequently is the data updated?)
- Other systems that this system is dependent on (this could be the number of feeder systems)

How do you uncover this information? You will often resort to sampling the data in the database and then manually reviewing it. You can also interview end users who work in the business process to obtain their views on and experience with the data.

What do you do with this information? Use the same approach as in the previous step. Construct a table in which the rows are the bulleted criteria and the columns are the systems. The table entry is 1 through 5 based on 1 – poor and 5 – good. You can also construct a spider chart for the systems.

STEP 3: DETERMINE YOUR KNOWLEDGE MANAGEMENT STRATEGY

With the previous two steps you now have information to serve as input to developing your strategy for knowledge management. You might also research what methods and tools you might consider for use in operation. These are discussed in Chapter 13.

Start with defining alternative strategies. Here are some examples that were used in three of the organizations used as examples:

- *Do nothing*. This is an important alternative and can be used to compare with any other strategy. What is the price of doing nothing? What opportunities are lost?
- *Lessons learned data base*. This is discussed in detail later through an example. The database consists of specific guidelines for how to do the work in the business process.
- *Go for a data mart*. You will extract data from current systems and create a database.
- Embark on a full-fledged data warehousing project.

Some comments are appropriate here. A data mart strategy offers a shorter route to benefits. Furthermore, if the firm has no experience with data extraction, data mining, and in-depth statistical analysis, this is a good way to get experience. The data warehouse alternative may be necessary if there is sufficient competitive pressure. You can also combine the lessons learned database strategy with other strategies.

Keep in mind that you must make these strategies specific to the business area that you want to address. Thus, you could have several strategies involving data marts or lessons learned. For the Roberts Agency, there were potential data marts for bus operations, finance, and bus maintenance.

After defining the potential strategies, you can now begin the comparative analysis of these. Build a list of evaluation criteria. Here is a list to give you a head start:

- Impact of the strategy on IT resources. Each strategy requires a different level of commitment from the IT organization.
- Impact of the strategy on business departments. For the lessons learned database, there must be intense department involvement. The same is true for the data warehousing. For the data mart, there is less effort than the data warehouse.
- Impact of the strategy on the business processes. When you devote resources to creating a data warehouse, you almost inevitably have to pull people from the business process.
- Potential risk of failure for the strategy. If the strategy fails, what are you left with and what can be salvaged?
- Potential short-term benefit. Take a time horizon of 1 to 2 years here.
- Potential long-term benefit. Use a 3 to 5 year time horizon.
- Availability of skills and resources to carry out the strategy. What technical

and analytical skills are available? What business skills and knowledge can you draw upon?

- Estimated elapsed time before benefits start to be achieved. Knowledge management involves a substantial start-up effort before there are any benefits.
- Estimated cost of implementation up until the time benefits are first realized. This is your estimated exposure.
- Estimated potential long-term cost of the strategy. This is almost impossible to estimate with accuracy.
- Potential impact of technology on the strategy. As technology improves, what is the impact of this on the strategy? Maybe, you should wait if there are going to be major advances.

You can use the table approach here. The rows are the criteria and the columns are the alternative strategies. The table entries are comments about the criteria applied to the strategies.

After this analysis is completed, you are ready to present the results to management. Make sure that you keep the alternative of doing nothing out in the open at all times. Do not appear as the backer of an aggressive strategy. Prior to any general management meeting, present the results to business departments and IT. The business department will have to support a strategy involving them in terms of their commitment to the effort and the estimated benefits. IT will have to be involved because they will have to support the feasibility of implementation.

EXAMPLE: THE LESSONS LEARNED KNOWLEDGE BASE

One of most useful and easier knowledge management applications to establish is a knowledge base of lessons learned. *Lessons learned* consist of the experience gathered in how to perform work in a business process. Lessons learned are valuable in that they can do the following:

- Reduce the learning curve of new employees to the business process.
- Eliminate errors made by the business staff in interpreting procedures to handle unusual business transactions.
- Facilitate the speed and throughput of business work by providing more accessible guidance to help in how to do the work.
- Provide guidance to the IT and business staff in understanding requirements and in making improvements.
- Provide a cumulative base of knowledge of the work.

Lessons learned can be gathered for almost all business processes.

Table IV
Basic Lessons Learned Database

Data element	Comment
ID	Identifier or number of lesson learned
Title	Brief descriptive title
Status	Submitted, Active, Replaced, Pending (under review)
Type	Category of lesson learned
Linkage to business process	Depends on the specific database (see chapter text)
Description	
Business process	Process to which the lesson learned applies
Situation that it applies to	This must be very specific so that the lesson learned is not misapplied
Date created	
Submitted by	
Specific guidelines or actions of the lesson learned	Step-by-step instructions
Benefits if applied	
Related lessons learned	
Expert or person to contact	Potential contact if user has questions

A lessons learned database is dynamic and is updated as people learn more when they apply the existing lessons learned. The database must reflect the cumulative experience. Here are the data elements for the basic lessons learned database (Table IV). These set up a lesson learned. Table V provides the data elements for adding on additional experience and guidelines. In practice, all lessons learned are generated or submitted to a *knowledge coordinator* who reviews the material and answers the following questions:

Table V
Additional Data Elements

Data element	Comment
ID of lesson learned	For reference
ID within lesson learned	For indexing
Title	
Status	Same as lesson learned
Description	
Experience in application	Text discussion of what happened
Suggestions	Hints for expanding or adding to lesson learned
Date created	
Created by	
Additional comments	

- Is the lesson learned too specific? Can it be generalized?
- Does the lesson learned indicate detailed actions to take?
- Could the lesson learned be misapplied?
- How long and under what conditions does the lesson learned apply?
- Does the lesson learned adhere to company policy?
- Is the lesson learned easy to apply?
- Are the benefits from applying the lesson learned evident?
- Without the lesson learned, what would people likely do?

A lesson learned database was established for the Roberts Agency. Because this involved bus drivers, there ended up being three separate databases. One database was for buses and equipment. This information provided guidelines on what to do if the driver encountered specific mechanical conditions. The second database included specific customers who were problems or required special assistance. The third database referred to the streets and intersections. Items in this database include dangerous intersections, tight corners for turning, and information on specific bus stops.

There was little hope given by management that these databases would be used. To their surprise, within 2 weeks the drivers had submitted more than 300 lessons learned. These were set up on intranet web pages. To date, there are more than 700 lessons learned with more than 250 additional comments and observations.

A critical success factor in using a lessons learned knowledge base is to determine how people can gain access to what they need fast. If you just hand someone a book and they have to thumb through it to find the lesson learned that applies to their situation, it will not be used. You have to be clever in trying to figure out how to provide access and indexing. Here are some examples:

- For the Roberts Agency, the approach was to link the lesson learned database for the streets and intersection to a Geographic Information System (GIS). You just click on the site and the lesson learned appears.
- For the database of people, the driver can search on the basis of a physical attribute or location that the person has encountered.
- For the database of the vehicles, the method was to search on the part of the vehicle.
- For general business processes, you can search on the basis of the transaction and the step within the transaction.

Think knowledge access through before you start gathering lessons learned. You also have to make it easy for users to add lessons learned and to access the database. For the Roberts Agency, the drivers can access the knowledge base from home through the Internet using a firewall.

As you gather these lessons learned, you must periodically go through the lessons learned and scrub these. You will take the comments and additions and then restructure the lesson learned. You will also create more lessons learned.

The value of lessons learned does not announce itself. To management, the benefit may seem obscure. Here are some measurement suggestions:

- Produce statistical reports on the number of lessons learned and additions. Also, indicate the number of accesses or hits on the database. You might even display the number of visits to the web pages when you set it up.
- Conduct surveys of the audience who is using the lessons learned knowledge base. At the Roberts Agency, the following questions were used:
  —How often have you accessed the lessons learned knowledge base in the past 6 months?
  —How has the knowledge base helped you?
  —What additional lessons learned would you like to see added?
  —What would you like to see changed?
  —If there was no knowledge base, what would you do?

This last question tends to demonstrate the true value of the knowledge base. In the case of the Roberts Agency, the drivers indicated that there were fewer accidents. The downtime for buses was reduced. New drivers believed that they had received the most value. Experienced drivers liked being able to share experiences.

WHAT CAN GO WRONG?

- For the lessons learned database, you can encounter resistance on several grounds. First, management may not see the value of it. Gather lessons learned manually bottom up to show that it is feasible. This also establishes a structure for lessons learned to which people can relate. Another point is that senior people may not want to share the information. They might believe that the information gives them an edge. To address this, you can offer anonymity to the person so that they will not be bothered. You can also apply peer pressure by collecting lessons learned from others.
- Some firms start out with a goal that is too ambitions. They also may only have fuzzy applications in mind. In either case, the project is jeopardized. Keep the initial goals finite and limited. Have very specific applications in mind.
- Cost justification is often quite difficult for knowledge management applications. Here are some suggestions. Take one of the questions that the knowledge management application is supposed to address. Make sure you pick one that people will recognize as important. Then go through the steps to either manually or through systems generate an answer. Estimate the cost

of this effort. This gives you a base. Now prepare a list of similar questions. Rather quickly, you will be able to show that the setup of the knowledge base is justified if you only use it for a few questions.

REDUCED SCHEDULE AND COST APPROACH

There are two basic suggestions to a reduced schedule and cost approach. One is to start with the lessons learned knowledge base. This is gathered manually and requires little programming and setup effort. It also does not require extensive programming to extract information from existing files and databases. The second suggestion is to first establish a data mart from several existing databases. This requires less effort than a data warehouse and can demonstrate benefits.

Once you have something established, move on it quickly. That is, start generating analysis and reports for both management and end users. Target end users because they will best be able to explain to management the value in their tactical operations and business processes.

Another idea is to have the business department staff heavily involved in the development of the knowledge database. If they are involved, they have a better understanding of what is occurring. Through involvement, they will also become more committed to it.

EXAMPLES

Atlas Bank attempted to establish a data warehouse for credit card application processing in the mid-1990s. They made the basic mistake of starting without proper analysis. The project died for several reasons. First, the data in the credit bureau was insufficient for analysis. Second, the hardware technology was too primitive. Third, the applications were so old that they could not feed the warehouse. A few years later, the bank tried again. The need was compelling for competitive reasons. The technology and credit bureaus had improved. Several old systems had been replaced. Most important, the bank management scaled back the objective to a data mart.

A midwestern utility had a great deal of customer data. The utility industry was being deregulated. The utility could expand into new areas. They thought of offering a credit card to customers. Without initial analysis, they constructed a data warehouse. When all was said and done, it turned out that only a small subset of customers qualified and were interested. Instead of trying to find other applications, the system was dropped.

LESSONS LEARNED

- You can try to embark on a grandiose data warehouse. You will likely fail. Concentrate instead on a data mart.
- Although data marts are very attractive in the short term in terms of giving tactical benefits, they can make life more difficult in the longer term. For example, Secour Retailing built three data marts. The data elements of these overlapped. When they wanted to create a data warehouse, they basically had to start over.
- What if you perform the analysis and find that there is little interest in knowledge management? You know that the applications are there, but management seems more interested in tactical, short-term results. Do not be discouraged. First, technology is on your side. As it improves and more tools emerge, the cost of implementation will drop. Second, the existing applications are bound to be improved or replaced. This will also reduce implementation cost. Third, there is an increasing awareness of the value of information and knowledge. Support will grow in the future. Fourth, you can use what you have learned in the analysis to work for improvements in the current systems and business processes.

SUMMARY

Knowledge management was virtually impossible in the late 1980s due to limited technology and the fact that so many of the systems were legacy systems with old file structures. The modernization of application software based on database management systems and fourth-generation languages, along with the radical improvements in hardware, networking, and system software, have made data marts and data warehousing feasible. Knowledge management software is being built into the off-the-shelf data management tools. For example, OLAP tools are now part of many database products.

WHAT TO DO NEXT

- Look around in your organization for potential knowledge management applications. Try to find out what people are doing now for data analysis to address management questions.
- Select an important business process. Assume that information was available on the process going back over time and with great detail on

transactions and work. What questions could you answer with such a database? What value would the answers be in terms of carrying out specific actions?

- Does your organization take advantage of lessons learned? Most organizations do not. Is there any effort to collect information from senior employees to use in training with junior employees? Identify an opportunity for a lessons learned knowledge base. What would be the steps in setting this up? What are some benefits that would accrue if this existed?

Chapter 13

Analyze Knowledge Management Technologies and Services

INTRODUCTION

Chapter 12 indicates the early steps you must take to make decisions and implement a knowledge management application. In this chapter, categories of tools are reviewed. The individual products that fit in each category change and evolve over time. Here the attention is on purpose and general features and capabilities.

DATA PREPARATION

The preparation and clean up of data before it enters a data mart or data warehouse is a major undertaking that frequently is underestimated. It is important to bear this in mind as specific steps are examined. In each step, you should work with a sample of data to determine the extent of the problems and to learn how to use the tool proficiently.

The tools must perform the following functions:

- Modify or transform data in one format to another. This is needed for databases on the same and different hardware platforms.
- Perform manipulation and calculations of data based on business rules. An example might be the validation of the city field based on ZIP code.
- Combine records prior to being loaded into the data warehouse.
- Work with the metadata repository software to provide information on modifications, formats, and so on.

Software that performs these functions is offered by several vendors. These vendors take somewhat different approaches to doing the work. Some generate code in either COBOL or a fourth-generation language based on source and target data definitions and transformation rules specified by the programmer or analyst. These are useful in supporting data conversion. You can usually add your own programs and link them in using the tool. You will be generating many programs for a large data warehouse with many sources. Thus, analyzing the correctness of the code is often much more time consuming than the initial generation of the code.

Another tool uses a trigger or audit log to determine changes in one data source and then to extract and manipulate the data so that the data warehouse can be updated. This is easiest if the source is a relational database. Otherwise, you may have to write custom code (API-Application Program Interface) to do this.

The capture of data at periodic intervals from source files and databases is the purpose of some tools that create data marts. Again, this is easiest if the source is a standard relational database. Then you would generate a database query and add data modification rules.

Some of the features of systems that support this extraction and update process include the following:

- Support of access to sequential files in legacy systems
- Use of proprietary data languages to ease the programming
- Dependence on a metadata dictionary
- Data dictionaries
- Substantial auditing and reporting capabilities
- Scheduling software for extraction and updating

METADATA

Metadata consists of information related to data in a data warehouse. Without metadata, it would be impossible to perform updates and analysis. Metadata support is handled through software called the *metadata repository*. Included in the repository are the following:

- Description of all warehouse data components
- Location of all components
- Definitions, structure, and content of the data warehouse and views of data for end users
- Sources of the data
- Rules on how the data warehouse is filled or populated. These include transformation and integration rules, how the operational files and databases map into the data warehouse, and algorithms to manipulate the data.

- Rules on how data are delivered to end users.
- History of data warehouse operations (updates, versions, restores, etc.) and an audit trail
- Security access and authorization
- Metrics to measure warehouse use and performance

With many vendors doing their own thing, it became important to establish standards. The Metadata Interchange Standard defines metamodel standards.

The metadata repository provides software tools for updating and managing the metadata. It provides a simpler and more direct way to handle requests and perform analysis. The repository provides data administration functions and makes development of applications to use the data easier.

How does this all work? You first acquire a metadata repository software tool. Typically, the tool will be a client-server-based tool and will support the import of information from Computer-Aided Software Engineering (CASE) tools. This process is dynamic in that you will want to add more data sources to the data warehouse. Frequently, you will encounter problems with combining the external data into the warehouse even using CASE tools. Some of the problems are inconsistency of data names and meaning, unknown data quality, missing data, and inconsistent formats.

ON-LINE ANALYTICAL PROCESSING

On-Line Analytical Processing (OLAP) tools can be employed with data marts, data warehouses, and relational databases. Many of the database vendors are now offering OLAP tools as an option. The need for OLAP is based on the fact that the data to be analyzed is multidimensional. Examples are sales analysis, market analysis, and financial forecasting. A user wants to employ the OLAP tool to extract from the huge databases and produce summary information. A typical output of an OLAP tool is a spreadsheet in multiple dimensions. In traditional SQL (Structured Query Language) queries, a user might want to know about sales of a product over time. OLAP supports much more complex queries. One could ask for sales of product families by product, across several years, in specific regions, and sold through specific channels.

If you can imagine a three-dimensional cube then you can see the challenge that OLAP faces. If you assume that the dimensions are the regions, products, and time periods, then the cell entries are sales. Not all cells will have entries. This means that the OLAP tool must deal with *sparse data*. Many cells will be empty.

E.F. Codd, the founder of relational database technology, formulated a series of requirements for OLAP software. These are as follows:

- A business user should be provided with a multidimensional view that corresponds to the business situation and is easy to understand and use.
- The complex technology in the OLAP tool should be transparent to the user.
- The system should only access data that is needed by the query. It should also be able to access external data sources.
- Regardless of the size of the query, the system should not experience performance degradation due to size.
- The tool should be able to be accessed by several users at the same time.
- All data dimensions should be equivalent in structure and operational capabilities.
- The system should be of the client-server type.
- The system should adjust to sparse tables and matrices.
- The system should support easy data manipulation by a user (point and click).
- There should be flexibility in reporting.
- There should be no limit to dimensions or the number of levels for the aggregation of data.
- The system should recognize and deal with data hierarchies.

Obviously, the OLAP tool should interface to standard database management systems, support an SQL interface, and provide for refreshing of data incrementally without having to repeat the entire operation.

As with other types of tools, there are different categories of tools. In one category MOLAP, (Multidimensional OLAP), the tool uses a specialized multi-dimensional database management system. This is useful in financial analysis where you would have to perform statistical time series analysis. Another category is ROLAP (Relational OLAP), which works directly with the relational database management system. This second category is more common.

OLAP product features include the following:

- Drill-down capability to move to more detailed levels of data aggregation
- Three-dimensional graphics
- Statistical charts and graphs
- Specific formats for financial reporting
- Security control that is dimension sensitive
- Integration with the database management system
- Precalculated summaries and Wizards
- Application development tools for OLAP applications
- Internet and intranet links and access

The widespread use of the Internet and the World Wide Web have created opportunities for OLAP tools. This area is still evolving. Some of the products support the following:

- Generation of OLAP reports and output as static HTML pages, providing only static access
- Support of interactive queries where the HTML request is transformed at the web server into a structured SQL or OLAP request
- Support of applications on the Web using Active Server Pages or Java

MODELS

You are given a set of data and are told that there is a pattern or model in the data. You worked with this problem in high school to detect the next number in a sequence of numbers. A major objective in data mining and data warehousing is to detect patterns and models in data. If you have historical data, you can define a *model* that describes or characterizes the data. You then will apply this model to new data to determine expected values and to predict what will happen. You analyze some data and determine that there is an underlying repetition of information. This sequence of events or data that occurs repeatedly is called a *pattern*. The model reflects what you are trying to do; a pattern stems from the data itself. In modeling in a data warehouse, you will define a function based on fields in the records of the database. Once you apply this *predictor* to the data, you obtain a prediction or predicted value.

There are many applications of models in data marts and data warehouses. One application is to determine which customers you want to reach out to for new products (acquisition). Another model is based on retention of customers. A third is based on expanding or extending new products to existing customers. A fourth is to acquire new customers. All these are based on models that you construct and then apply to the database.

Models find many applications in business. In banking, a model was created for evaluating credit applications for farmers for a major bank. Using the model, the loan officer collected data from the potential customer and fed it into the model. The model then indicated the credit worthiness of the customer. This worked great for more than a year, and the bank increased its marketshare by more than 20%. Then economic conditions changed. The model was no longer valid. However, the bank continued to use the same model and incurred a huge loss in bad loans. What went wrong? The model was not updated. Models have to reviewed often based on the current circumstances.

Models suffer from a variety of problems. One is that there is a shortage of information so that the prediction is not statistically sound. Another problem is that the database has missing elements. Missing data is a common problem in developing a model. To cope with these problems, you often will resort to sampling of the database. This is not a book on statistics; therefore, there is no space to go into the various types of sampling. However, it is significant to note that you should select modeling software that can carry out a variety of types of sampling: stratified sampling, cluster sampling, random sampling, and taking every xth record. When you sample, you incur an error and a variance of the error. The software should provide this analysis to help you to determine the accuracy of the model.

It is clear from this discussion that you will need statistical capability to analyze the data. Look for statistical tools that support histograms, classification, analysis of variance, correlation and regression, hypothesis testing, contingency tables, and nonlinear statistics.

You can differentiate among types of models. Descriptive models are those that give you a basic understanding of data. Based on sales data, for example, this includes aggregation of data, ratios, weighted data, and scaling. Examination models explore relationships among the results of descriptive modeling. They include statistical analysis, clustering, and associations. An example is to relate the range of products in a store to sales. You might come up with data indicating that for each x% increase in product spread, you might expect to get y% increase in sales. Using relationships, you can then make predictions. This third category includes pattern recognition.

ARTIFICIAL INTELLIGENCE

Artificial intelligence (AI) is concerned with developing models of behavior that affect a physical situation. There have been successful applications of AI in such fields as playing chess where an IBM computer with AI software was able to outperform chess masters. AI has been used in areas such as games, planning, logic, robotics, and natural language. Your word processor uses some aspects of AI to do spell and grammar checking.

A subset of AI is expert systems. *Expert systems* are computer-based systems that reflect what an expert would do when dealing with a situation. One example would be the loan approval example given previously in this chapter.

Given a data warehouse, it is natural to ask whether AI can be applied to the data and whether expert systems can be created from the data. This has its best fit in situations that are limited and well defined. An example is the analysis of a credit card application where the applicant's data in the application and a credit bureau report are employed to determine a score. If the score is low, the application is rejected. If it is high, it is accepted. In between the application is given to a human to make a judgment call. This is called *behavioral scoring* and is used frequently by banks and financial institutions. It is also being used by insurance companies to evaluate insurance applications in terms of risk.

DATA MINING

Data mining consists of software and a process for extracting information from very large databases to address specific business issues. In data mining, you analyze the existing data in the database to develop a model. For example, suppose that you are a large firm that leases cars and trucks. You could analyze the data to determine criteria under which a customer might turn their vehicle into a competitor and lease another vehicle from them. Data mining could help you determine who is most likely to do this. You could then implement a rule that gives customers who fit the data mining profile an attractive new lease before the old lease expires.

When you consider data mining, you should answer some basic questions:

- Are the rules easy to automate and apply? Or is there still a need for an expert?
- Can the rule generate either cost savings or revenue enhancement? Is it cost justified?
- Are the situations to which the rule is to be applied well defined? If there is fuzziness, then you are at risk.

Data mining software should support the following features:

- The technique should support clustering analysis on the data.
- The method should find links among different records in the database.
- It should be able to find unusual records that fall outside what is expected.
- The method should be able to create rules and then test these on new or different data.
- The method should be able to handle data with errors and missing data.

In addition, any software should be able to interface to a relational database, be scalable in performance, and support validation.

A commonly used model is a *decision tree*. In a decision tree, you walk down from the top of a tree. At each branch, you are faced with several choices. Based on the characteristics of the person or the situation, you take a branch and continue. When you reach the bottom of the tree, you have a recommended action. Decision trees are very common in a variety of industries to deal with well-defined situations where there are a limited number of variables.

Software that supports a decision tree will create a tree based on the data in the data warehouse. As you apply the software, you will find that missing and bad data affect the tree. The software should support you modifying the tree and then reapplying the software. The technology behind decision tree software comes from statistical analysis and AI.

Neural networks are another application of data mining. Neural network software is intended to provide thinking similar to what a human would do. An example of a neural network is in banking where a bank is attempting to determine if someone is committing fraud. In data mining, neural networks are really predictive models based on data mining of the data warehouse. It takes substantial expertise to design a neural network. A big problem is in presenting the network to an end user. It is often too complex for people to grasp.

A neural network consists of nodes and links. A node corresponds to a neuron in the human brain. A link connects two neurons. A simple application would take in several inputs such as age, number of years living in the same location, and income. These would be input nodes. These then are linked in a network to produce an output node that is a decision on credit. You create a neural network by feeding cases to the software that includes the predictor value. Each input node is given a different weight. You can manipulate the result by applying different weights. The neural network can learn in that you can keep feeding it cases. Neural networks for credit card fraud look at the behavior of the card use by the cardholder. For example, if the person has not traveled much and there suddenly appear a series of charges in a foreign country, the system raises a flag. Of course, you could do this manually. However, given the millions of credit card transactions, it would be impossible to handle this work manually.

DATA VISUALIZATION

The purpose of data visualization is to allow you to view information so that you can detect patterns in the data. If you stare at tables of numbers, you have a very low likelihood of seeing patterns. Thus, data visualization is a key to getting the most out of all the investment in the data warehouse or data mart and the tools discussed previously.

Data visualization software should allow you to perform the following:

• Compare data.
• Filter data to look at specific subsets (e.g., sales in various states).
• Move from the visualization back to the data source.
• Look at the graphs with different scales of detail.

Data visualization links to the other tools. When you do data mining, you would need the visualization tool to find a good place to start. As you do the data mining, you want to display the results visually.

How does data visualization work? The computer screen is only two-dimensional. You might have 10 dimensions to examine. You could work with them 2 dimensions at a time—a combinatorial nightmare! Instead, many visualization packages show each dimension on the horizontal axis. The vertical axis varies by each dimension. This reduces the situation back to two dimensions. Showing decision trees is a natural for data visualization. Summary data can be displayed in each box of the tree. As another example, you can overlay three-dimensional bar graphs over a geographic map.

COMBINING METHODS

In performing data analysis, you often must combine the different methods. We consider how OLAP, data mining, and visualization can be combined. Table I provides some comparisons of the three techniques. In this table, the column data definition refers to the data definition language (DDL) of the method, which is the data aspect of the tool. The next column is the manipulation language; which refers to the processes used. The last column is visualization of the information. Here you can see the differences among the three tools and the benefit of combining them when you are constructing decision support applications.

Table I
Characteristics of OLAP, Data Mining, and Visualization

Method	Data definition	Data manipulation	Visualization
OLAP	Dimensions, hierarchies	Aggregation, filtering, etc.	Drill-down type output
Data mining	Basic flat files	Decision trees, neural networks, etc.	Basic reports
Data visualization	Basic flat files	Basic capabilities	Charts, graphs

Suppose that you have an electronic commerce operation and want to do sales analysis. You would begin with OLAP to aggregate the data from sales of individual items. This is still very detailed so you might use data mining to cluster the OLAP results. What do these mean? Use data visualization. With the results of this step, you can return to OLAP and do more aggregation.

This discussion shows that when you are considering the tools for analyzing data in a data warehouse, you should evaluate all categories of tools, even if you do not initially foresee using all of them at the start. Also, it will be important to ensure that the tools work together. The last thing you need when you are doing this is dealing with either incompatible software tools or tools with entirely different interfaces. Because you will not be doing the analysis every day, there will be a relearning curve if there are substantial differences in the structure of the tools or the user interface.

EXAMPLE: ENTERPRISE RESOURCE PLANNING

The previous section of the book discussed Enterprise Resource Planning (ERP) software packages. ERP software allows you to collect massive amounts of data in a consistent framework. A major benefit that is put forward for ERP software is the ability to analyze all the data. Thus, it is natural to conceive and require a data warehouse that is fed by the ERP. ERP systems have been most successful at the transaction level. If you were to access the ERP data directly to perform analysis, the performance of the ERP software would most likely be impacted.

With the growing use of ERP, there are specific data analysis needs in specific industries. Examples are manufacturing, human resources, banking and financial services, accounting, marketing, sales, distribution, and competitive analysis. In evaluating software for these areas, you should consider some more specific evaluation issues than are discussed in Chapter 3.

- What is the audience of the software? Is it designed for a senior-level manager or for a highly trained staff analyst? Most are designed for the analyst.
- What is the range of analytical tools available?
- Does the software offer standard templates for different situations? This will save a lot of time and get you started earlier. The templates should also be able to be customized.
- Does the package support input from other sources beyond the ERP software? In particular, most companies have not replaced all their legacy systems with the ERP software; thus, the tools must still interface with legacy systems.
- Because it is unlikely that you will find a single package to perform all the analysis, you should ensure that the software can interface with other tools.

CONSULTANTS IN KNOWLEDGE MANAGEMENT

The discussion so far has centered on the software tools to support knowledge management. For many organizations, this is not sufficient. You also require some expert assistance in getting the software set up and in getting started. You may also need ongoing part-time expertise to refine and build new models.

Knowledge management consulting firms are somewhat of a different breed than standard systems consultants. First, due to the range of potential tasks they could perform, it is useful to consider hiring more than one group. For example, you might select one firm to help establish the data mart or data warehouse. Then you might pick a different one who has a proven track record of expertise in statistical analysis.

A second factor is that there is a wide difference in the tools within a category and how the tools are used. Whereas you can employ generalists in other IT areas, here you should consider firms that have expertise with the specific tool.

What are the benefits of using consultants in knowledge management? Here are some benefits from past projects:

- If the consultants work side by side with your business and IT staff, then the internal employees will get up to speed more quickly. There will be more opportunities for the transfer of expertise.
- If they have worked in the same industry as yours, the consultants can bring this specific knowledge and experience to suggest specific areas for data mining and OLAP. This can save time and produce results faster.
- Consultants can assist in getting some quick analysis results with even a sample of the data. This can assist you in dealing with management who may be getting more impatient due to the lengthy start-up time of the database.

Consultant evaluation was addressed generally in the last part of the book. Here are some suggestions to think about when evaluating knowledge management consultants:

- Contact the software vendor for recommended consultants who know their products. If the vendor is reluctant to do this, then there may be issues with the product.
- Narrow down the list to those who possess experience in your industry segment in your country. If you are the first one and there is no match for a given product, you risk becoming a pioneer.
- When checking out references supplied by the consultant, concentrate on the following:
  - Specific activities performed by internal staff versus consultants (do you have the right mix of skills?)
  - The exact tools and methods that were used (for a close fit and to ensure that you have all the tools needed)
  - The characteristics of the data warehouse created (to ensure a valid comparison)
  - Elapsed time it took to show results (match it against your plan)
  - Problems, surprises, and lessons learned encountered
  - What tasks did the consultant and client do?
- As with major software packages such as ERP, you should perform the consultant evaluation at the same time you are selecting the tools.

EVALUATION OF PRODUCTS

You could begin with reading books and articles and collecting feature lists for tools. Then you could use this information as the basis for the evaluation. Beware though that this traditional approach has serious shortcomings here. Here are observations and guidelines:

- The vendors are merging and being acquired at a rapid rate.
- Although there are some leading firms, there is no defined set of dominant firms as yet. This makes your selection even more important. If you select a product from a vendor who is acquired, then the new firm may set a different direction for the product—leaving you high and dry.
- There are emerging standards, but as yet there is no uniform interface between products. This makes the evaluation of interfaces between products a must in the selection.
- Many software products are not really affected internally by hardware advances because they depend on the operating system. This is not true for some of the categories here. As 64-bit computing gains popularity, along with increased support for multiprocessing, the firms have the opportunity to improve their software greatly and develop new products. In other words, the leader today may not be the leader tomorrow.
- Some of the better products come from small firms. These firms offer less product support than you would traditionally expect.
- When companies use these products, they tend to value the support and interfaces as much as the internal capabilities of the product.

Over and above the products, you must consider the audience for the tools. Who is going to perform the analysis? Where will it be based in the organization? Some alternatives are as follows:

- *Marketing.* This is often where knowledge management begins because it is often here that the data is most readily available and the issues well defined (e.g., retailing). However, marketing has its own local interests.
- *Planning.* Although this group can pose and address large-scale issues, they are reluctant to address tactical issues, which are the purview of the line organizations.
- *IT.* This seems to be an obvious choice, but many of these tools are end-user oriented. Offering this type of service would change and challenge the normal IT roles.

None of the above is a perfect solution. So what do you do? Identify three or four organizations that could use the tools. Involve these groups in the selection of the tools to get them involved. In Chapter 14, this idea is extended to implementation.

WHAT CAN GO WRONG?

- Management gets excited about the potential of knowledge management and adds more resources to expedite the work. This creates even more pressure and problems. Not only do you have the problem of figuring out what people will do, but also management's expectations are raised. The underlying problem is that it takes substantial elapsed time to design and implement the database.
- If you miss one of the software tools needed, then going back later could slow down the overall effort. The most common problem is that the more exotic tools are viewed as not necessary. Even if they are not purchased initially, they should be selected.
- You spend a great deal of effort and establish the database. Then you find out from using the tools that the data is in worse shape than was estimated. To head this off, carry out sampling with the source data.

REDUCED SCHEDULE AND COST APPROACH

To ensure that your tool architecture is solid, select the tools at the start. To save money and give yourself flexibility, do not commit to buying the tools until shortly before they will be needed. Although the interface development and implementation of the tools take time and are not easily reduced, you can concentrate on reducing the learning time for staff working with the tools. Furthermore, form a working committee of people from different departments so that they can share experiences, and establish a lessons learned database just for knowledge management and the tools.

EXAMPLES

Secour Retailing implemented a data mart. The software tools were selected and acquired one by one. IT management did not want to overwhelm the business or IT staff. The first tool was an OLAP tool. After some initial analysis, which did not reveal anything startling, they began to search for other tools. Management grew impatient and brought in consultants. The entire effort was taken away from IT and placed in the planning area.

Vision Insurance wanted to move into knowledge management but lacked the resources for a full-fledged effort. Instead, management directed the creation of a customer database from several legacy systems. Without tools, the only thing that could be done was to generate successive SQL queries and then store the data in spreadsheets. This approach continued for some months. In the meantime, two major competitors launched new products that Vision Insurance had conceived of first. The IT staff was pulled off the project and assigned to enhance a legacy system to handle the new product. This reveals the importance of management commitment and an implementation plan.

LESSONS LEARNED

- Carry out training in the software tools just before they are used in a test mode. If you do the training too early, the techniques will be forgotten.
- Use the sample data that you collected to check validity and quality as a base for training in the use of the software analysis tools.
- In working with different departments, make sure that they have more than one person who works with the software tools. People with these capabilities are in demand. If the one person leaves, then the knowledge management experience goes with him or her and the tools are no longer used. This is such a serious concern that you should insist that departments have several people trained and proficient in the use of the tools.
- Templates and models should be documented. This is not only beneficial in case someone else must use the tools, but the template and model gives credibility to the results presented to management.

SUMMARY

In the perfect world, because knowledge management is evolving in terms of tools and being affected by hardware and network advances, a suggestion would be to wait and just work on the lessons learned knowledge base. However, it is not that simple. The questions that line organizations and management want to answer are really vital to the company. Competitors are pursuing these tools. Moreover, firms that invested millions of dollars in ERP software want more benefits through data analysis. Knowledge management is an exciting area because it addresses one of the golden promises of data processing from the 1950s—the availability and capability to analyze information—which gave rise to the phrase management information systems.

WHAT TO DO NEXT

- Begin to identify magazines and sources on the web that deal with knowledge management. Identify one firm in each of the tool areas discussed in this chapter and track the change and evolution of the products.

- If you do not have data marts or data warehouses, you might consider the potential audience in various departments that could use the tools. See if they have some interest.
- If you have implemented a data mart or data warehouse, then you might consider measuring the effectiveness of the tools through interviews or a survey. Also, look through the literature to see how these products have fared in capabilities over time.
- Find articles that give surveys of knowledge management tools in one category (e.g., OLAP, data mining). Then go into the archives and find a survey article for the same area 3 years ago. Compare the two articles and you will get some insight into how the technology and tools have advanced.

Chapter 14

Implement Knowledge Management

INTRODUCTION

In chapter 12, a series of steps is given for implementing knowledge management. The following steps have been completed:

- Identify knowledge management opportunities (Chapter 12).
- Assess the current state of the systems and data (Chapter 12).
- Determine your knowledge management strategy (Chapter 12).
- Select the appropriate methods and tools (Chapter 13).

There is sometimes a tendency to plunge into implementation after doing all this work. Resist this temptation. Implementation of knowledge management may be more difficult and complex than you think. Here are the implementation steps:

- *Step 1: Measure what is currently occurring.* This enables you to establish a baseline for comparison later.
- *Step 2: Determine your implementation strategy.* This is different from your overall knowledge management strategy. Implementation strategy centers on what the sequencing of actions will be and the how the organization and support can be prepared and structured for knowledge management.
- *Step 3: Define the implementation plan.* This is a detailed project plan. Knowledge management involves organization, technology, systems, and information. You will be more successful with a formal project plan.
- *Step 4: Establish the infrastructure for knowledge management.* This includes the hardware, system software, and network components and systems so that you can install the knowledge management software on top of this infrastructure.

- *Step 5: Implement the Data Mart and Knowledge Management Tools.* You have identified what software you require in Chapter 13. Now you have to acquire it, install it, test it, and train people in it.
- *Step 6: Complete the first knowledge management application.* It is assumed that this will be a data mart. However, there are two other possibilities—a lessons learned knowledge base and a full-fledged data warehouse. All three are examined here.
- *Step 7: Measure the results of the knowledge management application.* Do not assume that because everyone is happy that there will not be recurring questions. After all, knowledge management is expensive to implement. People will question the expense if they do not realize the benefits. Here you measure the results of the implementation and compare it with the measurements of Step 1.

Between Steps 6 and 7 and after Step 7, you will be expanding the number of users as well as creating new applications. You will manage and support knowledge management.

These topics are addressed in this chapter. Many books describe knowledge management but few go into the nitty gritty of implementation. Maybe that is why so many knowledge management projects start but so few finish successfully.

STEPS IN IMPLEMENTING KNOWLEDGE MANAGEMENT

STEP 1: MEASURE BEFORE YOU START

You already have a lot of information about the current situation. In this step, you will structure the information to develop a score card. This is really a process and infrastructure score card that evaluates where you are now. This will always be beneficial because you can measure again in the future and compare it with this past situation. If you do not measure, you just have word of mouth and subjective opinions—not much to go on if the value of knowledge management is attacked. Why be so defensive? Because knowledge management is so new to an organization. It is not a traditional application. It may require specialized hardware, software, and software tools. It is expensive and requires a long lead time.

Figure 14.1 provides a list of factors in your measurement approach. For each, we discuss where you get the measurement data, how to measure, and how to develop a consensus on the information. Keep in mind that this information will be used in the definition of the implementation strategy and in the development of the implementation plan.

1.   Descriptive information
      1.1.   Name of opportunity
      1.2.   Description
      1.3.   Organizations involved
2.   Current business situation
      2.1.   How work is done today
      2.2.   Problems with current approach
             2.2.1. Data problems
             2.2.2. Difficulty getting information
             2.2.3. Lack of tools for analysis
             2.2.4. Lack of time and resources
             2.2.5. Skill sets of current staff
      2.3.   Impact of not being able, to perform analysis
             2.3.1. Department impact
             2.3.2. Management impact
             2.3.3. Customer impact
             2.3.4. Supplier impact
             2.3.5. IT impact
3.   What competitors and other firms are doing
      3.1.   Benefits that they perceive
      3.2.   Tools and methods used
      3.3.   Infrastructure used
      3.4.   Lessons learned
4.   Knowledge management approach
      4.1.   Knowledge management strategy
      4.2.   Expected benefits and impacts from knowledge management
             application
             4.2.1. Business department
             4.2.2. Company and management
             4.2.3. Customers
             4.2.4. Suppliers
             4.2.5. IT
      4.3.   Infrastructure
             4.3.1. Hardware requirements
             4.3.2. System software (operating system, utilities, relational database
                    management, fourth-generation language, etc.) requirements
             4.3.3. Network (hardware, cabling, circuits, etc.)
      4.4.   Knowledge management tools
             4.4.1. Data preparation, extraction
             4.4.2. Data cleanup
             4.4.3. Metadata
             4.4.4. OLAP
             4.4.5. Models
             4.4.6. Artificial intelligence
             4.4.7. Data mining
             4.4.8. Data visualization

Figure 14.1   Measurement factors for knowledge management.

Current Business Situation

Briefly describe what work is done to analyze data in the situation. Then go into how that work is being performed. You might want to use a standard flowchart. Do not go into too much detail. Keep it to 10 to 15 steps.

Identify the problems with the current approach. Five categories of problems have been identified to get you started. Data problems include data incompatibility, poor data quality and accuracy, age and lack of timeliness of information, and missing data.

Difficulty in getting information refers to the many steps required to obtain the information. For example, it may be necessary to obtain or generate reports from several legacy systems. Then you have to interpret and modify the data based on knowledge and reenter the data into a spreadsheet. Summarize these problems in a list.

Lack of tools refers to the inadequacy of the software being used. In many instances, it is just a spreadsheet or simple personal computer (PC) database management system. Indicate what capabilities are missing. The people who are doing this may have many other commitments. This may be something they get around to when they can. Related to these are the skill sets and knowledge of the staff doing this work. Perhaps they lack statistical expertise and do not know how to do data analysis.

There are effects of not being able to perform the analysis, doing incomplete analysis, or not providing information and analysis results in a timely manner. Identify the effects by who is impacted. Departments, management, IT, customers, and suppliers have been listed to start. You can also discuss what will happen in terms of impact if the situation continues. What deterioration will occur?

Competition and External Information

If you contact other firms or gather information from magazines and the Web, then you might be able to pin down the technology and tools that they use. Articles and other sources may also indicate the benefits received and lessons learned. The software suppliers of knowledge management software tools are always searching for success stories that they can publicize. These articles help push their product as well as boost knowledge management.

Knowledge Management Approach

Summarize your knowledge management strategy in a list of bulleted items. It is possible that there is no overall strategy. Perhaps it is a tactical strategy to identify and build a customer base, or maybe it is strategic to learn more information about product sales. This will be useful later as a reference. In earlier steps involving the analysis, you defined benefits of the knowledge management application. Now make a list of benefits in terms of the audiences that are affected by the current method.

In terms of infrastructure, identify the hardware, network, and system software requirements for the knowledge management application. If none are required, indicate this. Infrastructure is discussed in more detail later in this chapter. Keep in mind that knowledge management generally requires fast processing and support for vast amounts of information. Thus, you will probably have to acquire additional hardware. It often makes sense to acquire a separate system for knowledge management applications. Otherwise, the knowledge management application can drag down the overall performance of the system. Along with the infrastructure, identify the tools that you have selected from Chapter 13.

STEP 2: DETERMINE YOUR IMPLEMENTATION STRATEGY

How should you implement knowledge management? The ingredients of an implementation strategy include the following:

- Hardware, system software, and network
- Knowledge management tools
- Interfaces to existing systems and data cleanup and extraction
- Organizational steps, such as setup, training, and preparation, to perform the new functions
- Consultants and vendors
- Business process, procedures, and policies to support knowledge management

In what order should you do work with these items? Should you acquire hardware before you even have an application? This could lead to more political exposure. How long can you wait before dealing with organizational issues? What if the current people that would use the knowledge management tools are not the right ones for the new work? When should you hire consultants? To address these and other similar questions, you must have a strategy.

You should consider alternative strategies and then select one. For example, for Atlas Bank the following three potential strategies were identified:

- *Throw money at the application from the start.* Under this strategy, you would deploy the new technology and tools as soon as possible. Interface work and data cleanup and extraction would begin. Consultants would be brought in early as well. Organizational and business process issues and tasks would wait.

There are several shortcomings to this approach. First, there may be a reluctance to make this kind of financial commitment at the start of something that has not been proven in the organization. A second problem is that if you defer organization and process issues too long, the business may begin to question the project and lose interest.

• *Try to do all the work internally.* Under this alternative, you would concentrate on internal resources. Hardware and other infrastructure enhancements would wait. The use of consultants would also be deferred. Organization and process topics could be addressed, and some work could be done gradually on data extraction.

There are problems with this strategy. The internal staff lack the knowledge and experience of knowledge management. This will cause the work to take longer than it should. Consultants could help, but they are not available. The positive organizational and process work would not lead to immediate action and results because the other parts of the process would lag behind.

• *Be conservative at the start and then expand rapidly.* With this alternative, you would buy the minimum required hardware and system software to get you live with one application and make almost no network changes. You would employ consultants on a limited basis to get you started. The attention would be on data extraction and cleanup as well as on organization and process work.

The last alternative is preferable for several reasons. First, it defers as much spending as possible until the application benefits have been proven. Second, it concentrates work on the areas of likely risk — getting the data and establishing the new process and organization roles.

STEP 3: DEFINE THE IMPLEMENTATION PLAN

Here the highlights of developing the project plan are touched on. For more complete information on general project management the reader is referred to two related books by the authors (*Project Management for the 21st Century* and *Breakthrough Technology Project Management*). Both are published by Academic Press.

Action 1: Develop the Project Concept

The first thing to do is to develop a *project concept*. This contains the following ingredients:

• *Purpose of the project.* You should state both a business purpose and a technical purpose. For knowledge management, the business purpose might be to implement a knowledge management application and business process that addresses questions in a specific business area. Give some examples as well. The technical goal might be to establish the software, interfaces, and infrastructure.

• *Scope of the project.* What is included? Are changes to legacy systems to clean up data included? Are changes to the business department organization within the scope? The dimensions of the scope include technology, software, organization, and process.

- *Roles in the project.* Who will do what? Most important, what are the roles of the business department? What are they responsible for? What is the role of consultants? What tasks will IT handle? For a knowledge management application, the business department should take responsibility for developing training materials and procedures, ownership of the use of the knowledge management tools, gathering lessons learned, and supporting the conversion and interfaces with existing systems.
- *Issues.* These are problems and opportunities that you are likely to face in the project. This could be a long list. Figure 14.2 contains issues from past projects that should be helpful as a starting point.
- *General schedule of the project.* Most important are estimates of key milestones such as completion of infrastructure, interfaces, data cleanup, and the first application being ready.
- *Budget.* The overall budget should be estimated along with estimates on when funds will be needed.
- *Benefits.* With all the analysis that has been done previously, it is appropriate for the business department to step up to the table and estimate benefits that they think they can achieve.

- Certain managers resist knowledge management because they want to keep the data and analysis to themselves for reasons of power.
- There are unforeseen problems in interfacing several knowledge management tools.
- The business department is not very creative in thinking of ways to analyze the data.
- After initial use of the information, usage tapers off. There is no ongoing management drive for continued and expanded usage.
- There is no mechanism for taking the results of the data analysis and applying them to the standard business processes in the organization.
- There is new hardware available that could substantially improve performance.
- Programmers in the legacy systems keep changing the code of these systems, causing the interface into the data mart or data warehouse to blow up.
- Management uses the information and analysis to jump to major conclusions. Such conclusions are not warranted due to limitations on the data and statistical results.

Figure 14.2 Potential issues.

- *Change control*. This is crucial to knowledge management because of its fuzzy nature. You must implement a method for controlling changes to the purpose, scope, technology, and project activities.

The purpose of the project concept is to get issues and roles, as well as purpose and scope, on the table for discussion. You do not want to develop a detailed project plan unless you have a common vision of what people are supposed to do, what problems have to be overcome, and the scope of the project.

The project concept should be presented to management, the business department, and IT for analysis. If you have a consultant and the vendors on board, then they can also review this and provide some input. Spend a lot of time in the review and on revisions.

Action 2: Establish both Issues and Lessons Learned Databases

You have a list of issues from the project concept. Put these into a database for tracking. Also establish a database for lessons learned. Examples of lessons learned are provided in Figure 14.3. Use the database structures defined in Chapter 11. The issues and lessons learned will be related to the project tasks. Experience indicates that knowledge management project killers are the lack of resolution of issues, lack of awareness of issues and their severity, and failure to capture and use experience and lessons learned as guidelines for future work.

- It is better to involve many people part time in the project than fewer people full time.
- There must be more planning for hardware capacity.
- Due to resource limitations, there must be more analysis and justification of the extent of historical data that will be maintained.
- In the future, consultants must be more closely in touch with both IT and business staff.

Figure 14.3   Potential lessons learned.

Action 3: Construct the Knowledge Management Project Plan

How do you begin to develop a plan since that you have not done this before? First, you should define a list of resources. This list will serve as a resource pool to support future analysis. If you do not employ resources in your plan, it will revert to a drawing tool. Use off-the-shelf popular project management software that allows you to customize data elements for tasks, filters for the tasks, and views of the data.

Next, develop a high-level plan or template that provides summary-level tasks. To get you started, a sample template is shown in Figure 14.4. (Milestones are indicated by M:) Use this and perform the following:

- Identify who will be responsible for each general task.
- Assign these tasks and have people define more detailed tasks (down to 1 to 2 weeks) for the first 3 months of the project.
- Review the tasks with them.
- Have these people create dependencies among tasks and identify resources for each task.
- After reviewing this, have them estimate the duration and dates for the detailed tasks for the next 3 months. Also have them estimate the duration for the general tasks beyond three months.

These steps lead you to a project plan that the team has created under your direction. You can now set this plan as the baseline. You can also identify the tasks with risk and associate them with the issues in the issues database as discussed earlier. You can do the same with lessons learned.

```
        Project management
            Implementation strategy
            Project concept
            Issues database
            Lessons learned database
            Project plan
            M: Approval of roles
            M: Acknowledgment of issues
        Measure current business process
            M: Measurement completed
        Infrastructure
            Requirements
                Hardware
                System software
                Network
                M:  Requirements approved
            Procurement
            Installation and testing
                Hardware and system software
                Network
                Integrated testing
                M:  Infrastructure in operation
```

Figure 14.4   High-level template tasks and milestones. Milestones are indicated by "M":, and tasks included within others are indented.

Knowledge management software
    Analysis and selection
    Procurement
    Installation
    M: Tools installed and tested
    Training in the use of tools
    Pilot use of tools
    M: Trained staff
Technical interfaces
    Data capture and loading
    Requirements
    Technical approach
    M: Approval of technical approach
    Sampling of data
    Analysis of data
    M: Approval of data quality
Design and programming of interfaces
    Design
    Programming and unit testing
    Integrated testing
    Pilot test of interface
    Documentation of interfaces
    M: Programs completed
New business process
    Staff roles, responsibilities, and tasks
    Organization roles and responsibilities
    Policies
    Procedures
    Training materials
    Training
    Interface with standard business processes to get results implemented
    M: New process in place
Initial production period
    Monitoring of use
    Measurement of new business process
    M: Review of results and approval of recommendations

Figure 14.4   *(Continued)*

Here are some guidelines for presenting the project plan. Start with the summary plan that gives high-level summary tasks and milestones. Then zoom into the picture for the next 3 months. Next, show the list of issues. This will get their attention. Zoom back up to the general picture and highlight the issues in the GANTT chart.

Step 4: Establish the Knowledge Management Infrastructure

Establishing the knowledge management infrastructure has not been addressed in detail until now. What hardware do you need to acquire to run the knowledge management application? It is possible that you can get by initially using the current hardware. Note, however, that this will quickly become infeasible if you are successful. Knowledge management applications consume central processing unit (CPU), memory, and disk resources intensively. Plan ahead for acquiring a separate computer for knowledge management. You have a real advantage here in that hardware is dramatically improving and your knowledge management applications can take advantage of multiple CPU configurations.

Things are more complex if you have to implement a new database management system or fourth-generation language just to be able to install the data warehouse. Then you face a whole series of installation and learning tasks for a family of software products.

The network additions, if needed, should be routine. However, as an example, if you are running Novell and the new software only runs on Windows NT, then you must install Windows NT and establish interfaces with the Novell network.

Step 5: Implement the Data Mart and Knowledge Management Tools

You first might begin with defining and designing the database for the data mart. This will consume many hours because you have to determine whether the data elements can be usefully populated from the source systems. You will find that motivating people to do this work will be difficult. It is time consuming and there is a lot of detailed work. Moreover, the business staff who can do this are the most senior and valuable to the business department. They have to be separated from their normal work for substantial amounts of time. This is a lot to expect of line supervisors and managers.

In this step, you will analyze the interfaces required for loading the data from existing applications for the first time and provide updates on an ongoing basis.

This is likely to mean writing two sets of programs (one-time data load and ongoing updates). Testing these will consume time in production of the source systems and may be difficult to arrange and schedule.

In addition, you must write programs to transform and clean up the data—one time and on a recurring basis. Defining what and how to clean up the data will involve substantial business staff time in reviewing samples of the data from the current systems as well as the data definitions of the data mart.

Now run the interfaces and populate the data mart. What's next? Well, you have to test and verify that the information came over correctly and that any manipulation and transformations were carried out successfully. This requires a great deal of detailed review work by senior-level people. It will likely take several tests of this to actually set up the data mart.

In parallel to these efforts, the knowledge management software tools can be installed. People must be trained on test data. Unfortunately, the people who are to be trained in tools are the same ones that are trying to help you with setting up and validating the data. Concentrate on the OLAP tool first because it is usually the first one employed in analysis.

Once you have the tool and initial data mart established, you can perform larger-scale testing. The results may be counterintuitive. This will then lead backward to trace what happened. You will have to drill down with data mining to find out which records gave rise to the results. Then you will have to check these data.

As the data mart is being tested, you can begin to test the programs and the subsystem that extract changes in the source databases and files and update the data mart. You will be very concerned here with performance. If the extraction process takes too long, production performance for the source systems could be affected. Because this is not uncommon, you should plan on additional effort to tune and adjust the extraction programs.

In parallel to all this work, you will be working with the business department to establish the new business process to support the knowledge management application. Together you will be developing the following:

- Policies for the use and analysis of data
- Identification for how results are to be implemented into the standard business processes
- Procedures for doing the analysis
- Training materials for the process
- Security and access control rules
- Methods for assessing performance of the process

These will have to be verified through simulations and dry runs. There will be a need for updates and for filling in missing pieces.

STEP 6: CARRY OUT THE FIRST KNOWLEDGE MANAGEMENT APPLICATION

Finally, the pieces are in place to actually implement the new knowledge management process. The first analysis runs should be done with just a few people to shake down the procedures. Try out questions of the data and goals for which the answer is already known. Try questions that address different levels of detail. For example, if you are dealing with sales, then you might look at overall sales of a product as well as breakdown by region. Use these questions as cases for additional training later. Hold a postmortem meeting to identify lessons learned and issues.

After this, you should move in two directions. On the one hand, you can do more analysis with detailed questions. On the other hand, you should prepare a management presentation that you will deliver to management with business department staff. Hopefully, they will indicate the ease of doing the analysis. You can hope that they will say, "How could we ever have done without this tool?"

ESTABLISH THE LESSONS LEARNED KNOWLEDGE BASE

This implementation is different than that the data mart discussed previously. You begin to work on parallel fronts. You will be designing and programming the lessons learned database. This will be tested and documented. In addition, you will be having people identify lessons learned. For the Roberts Agency, there were literally hundreds of lessons learned gathered. You have to evaluate each one on the basis of the following:

- Is the lessons learned valid?
- What is the value of the lesson learned?
- To what situations does the lesson learned apply?
- Is the situation that gave rise to the lesson learned only temporary? If so, maybe it should be dropped.
- Are there specific suggestions and guidelines that apply to the lesson learned?
- Is one lesson learned related to another? If so, maybe they should be combined.
- Do any lessons learned contradict others? If so, you must resolve the conflicts.

Once the database of lessons learned is established, you will encounter new challenges. How do you get people to use them? For the Roberts Agency, the approach was to center attention on younger bus drivers with less experience. These people saw the benefits of the lessons learned faster. Some of the senior people believed that the lessons learned were common knowledge and so did not need them. Working bottom up, you use the junior people to encourage the senior ones to participate. You can also establish a bonus or incentive program for providing the most useful lessons learned. Here "useful" is defined by a committee of bus drivers or users.

After the database is in production, you can concentrate on expanding use and on scrubbing and improving the database. You can also review the lessons learned to see if they should be part of formal training, procedures, and policies.

MANAGEMENT AND SUPPORT OF KNOWLEDGE MANAGEMENT

Knowledge management is not a one-time affair. It requires constant monitoring and management. Here are some of the duties that are required:

- Monitor interfaces between the source systems and the data mart. This is important not only because of data contamination with a faulty interface, but also because you want to catch any error early.
- Monitor performance of the source systems and the data mart. Do this so that you have lead time if you have to acquire additional disk storage and so on.
- Encourage users to share lessons learned with each other. Otherwise, people will repeat the same mistakes because they are working in a vacuum.
- Establish a regular review process to screen questions of the data mart. This will force people to think things through. In particular, you want to encourage them to consider how results would be used.
- Monitor how results of the data analysis are being employed. Identify benefits.
- Look for ways of streamlining the process within departments and for sharing information among departments. As a person in one department acquires special expertise, try to make this available to others.
- Look for opportunities to expand the base of users as well as to implement and increase the use of more sophisticated tools.

STEP 7: MEASURE KNOWLEDGE MANAGEMENT APPLICATIONS

After the first application has been employed for a few months, it is time to revisit the process and measure knowledge management benefits and effects. A different checklist will be used (Figure 14.5). Some comments here:

1. Descriptive information
   1.1. Name of knowledge management application
   1.2. Description
   1.3. Organizations involved
2. Current business situation (with the new process)
   2.1. How work is done in the new process
   2.2. Status of original problems with the old approach
       2.2.1. Data problems
       2.2.2. Difficulty to get information
       2.2.3. Lack of tools for analysis
       2.2.4. Lack of time and resources
       2.2.5. Skill sets of current staff
3. General benefits and impacts of the new process
   3.1. Department level
   3.2. Management level
   3.3. Customer
   3.4. Supplier
   3.5. IT
4. Detailed evaluation of process
   4.1. System performance
       4.1.1. Response time
       4.1.2. Throughput
       4.1.3. Interfaces with legacy and other systems
       4.1.4. Data quality and completeness
       4.1.5. Availability of system
       4.1.6. Ease of use
   4.2. Business department
       4.2.1. Quality and amount of analysis
       4.2.2. Skills of staff
       4.2.3. Ability to capture and use lessons learned
       4.2.4. Involvement in and commitment to the process
   4.3. Consultants
       4.3.1. General contribution to the project
       4.3.2. Success, willingness to transfer knowledge to internal staff
       4.3.3. Technical quality of work
       4.3.4. Management of their own resources
   4.4. IT support
       4.4.1. Amount and extent of support
       4.4.2. Depth of technical knowledge
       4.4.3. Ability to capture and use lessons learned

Figure 14.5   Measurement of a knowledge management application.

5.   Lessons learned
    5.1. In the process
    5.2. The knowledge management strategy
    5.3. The implementation and its strategy and plan
    5.4. Hardware and network resource estimates
    5.5. Knowledge management software tools
    5.6. Understanding of the information

Figure 14.5 *(Continued)*

- You are evaluating an entire process. You are not evaluating only the software tools or the data. This means that you are considering everything from quality of use to performance of the system.
- In Section 2 of the checklist, you are revisiting the original issues and problems to see what has been solved and what remains. The problems that still remain may be affecting the benefits gained from the process.
- General benefits and impacts are discussed from the same perspectives as before in Section 3.
- Section 4 contains a more detailed evaluation of roles and results.
- Lessons learned from the project are captured and listed in Section 5.

You will obviously use this information in a management presentation. During the presentation, you should also present the results of meetings with staff involved in the process. A key question for them to answer is, "What would be happening now if there had not been a knowledge management application?"

What are potential actions that can be taken as a result of the measurement? Here are some from two of the example companies:

- Reprioritize the departments and applications based on experience. Focus on departments that have greater potential benefits and that are eager to get going.
- Consider upgrading the hardware and network infrastructure.
- Consider replacing, augmenting, or upgrading the knowledge management tools.
- Refine the project management template for future projects.

WHAT CAN GO WRONG?

- Unbelievable as it seems, some firms plunge in and start to build the data mart or data warehouse without a project plan. Why do they do this? They may see it as a different piece of work outside the normal IT arena. Some have said that if they developed and showed the plan, the project would be canceled.

- Beware of midimplementation paralysis. This occurs when you are in the midst of the project and are faced with seemingly insurmountable and intractable issues. Expect that this will happen, especially if you are creating a data warehouse. What should you do in anticipation? You should establish intermediate milestones. These milestones will give the project team a morale boost and show management that progress is being made. If paralysis does occur, concentrate on near-term tasks to achieve a milestone and work with management to resolve the issues.
- Somehow you picked the wrong tool, or during the project a much better one emerges. What do you do? Do not change horses in midstream. Keep going with what you have. Once it is in use, you can go back and review the tools as part of measurement and suggest replacement.

REDUCED SCHEDULE AND COST APPROACH

To cut costs early, defer as much of the hardware and system software purchases as possible. Also, defer the acquisition of the knowledge management tools. New versions keep emerging and you do not want to keep buying and upgrading to new releases. To control consulting costs, assign work to the consultants on a task-by-task basis. This will allow you better cost control.

To cut the elapsed time, do as much in parallel as possible. Get the business department staff heavily involved. By the time the software is installed, they should have the procedures and policies defined for the new business process, have them validated, and have questions and analysis topics lined up that should be addressed. They are also responsible for estimating and achieving the benefits. IT can only provide support; they cannot tell management what the benefits are.

EXAMPLES

The Roberts Agency implemented the lessons learned knowledge base. It was very successful almost from the start. Here are some of the critical success factors. First, lessons learned were gathered manually from the beginning and validated. This got bus drivers involved in the method of using lessons learned. It also showed management that they had an interest. Second, implement the system on an intranet that allows users to access the lessons learned from home. This proved very useful since they have very little free time once they show up for work.

Millenium Manufacturing considered a data warehouse but realized that they did not have the resources for implementation. They implemented two data marts. One was for customers. This one supported determining what customers ordered and helped structure future production. The second data mart was for installation in the field. This consisted of maintenance and replacement information. This information provided valuable feedback to improve production quality of manufacturing.

LESSONS LEARNED

- When defining the project concept, make sure that you give examples of items that are not within the scope. This will provide a clearer boundary of what is in and what is not.
- As you install knowledge management software tools, start giving demonstrations to managers. You are not trying to raise expectations. Instead, you are attempting to raise their level of understanding so that they can formulate better questions and understand results better.
- Be willing to change the project team as the project change progresses. The needs of the project change as you move from data interfaces to testing to actual use. Different skills are required. Also, the more people that you involve in the project, the more commitment and understanding you will get.

SUMMARY

Implementation is where the rubber meets the road. Frequently, too much energy is expended in the analysis prior to implementation. Then there is a desperate rush to implement—this is a big mistake. Approach implementation step by step. Several benefits to this approach are that you are more likely to gain understanding and commitment and you are going to have an easier time later when dealing with problems and issues.

WHAT TO DO NEXT

You can do a lot of preliminary work prior to the implementation:

- Develop your project template for knowledge management. Circulate a draft of the template to others to get reaction and feedback. Put it on the network or on an internal web page for people to view.
- Define a list of issues beginning with those in Figure 14.2. Relate these to the template.
- Construct the issues database and develop the data elements for each issue.

Part V

Electronic Commerce and Implementation Issues

Chapter 15

Implementing Electronic Commerce

INTRODUCTION

Electronic commerce (e-commerce) means transacting business transactions electronically. There are two types of e-commerce: business to customer and business to business. In the case of business to customer, the focus is on getting a brand presence and building a customer relationship. For business to business, although the customer relationship is as important, another goal is to streamline all business processes related to e-commerce transactions.

E-commerce includes taking an order over the internet and then printing it and faxing it to the warehouse for shipping and to accounting for processing. You can see that e-commerce as commonly defined is wide in scope. That is why there is another term that is often used.

Electronic business (e-business) means that all steps in doing the transaction are handled electronically. E-business is definitely superior to general e-commerce in that it produces greater efficiency and improved customer service. Although both e-commerce and e-business are addressed here, the bulk of the attention is on e-business. For the sake of simplicity, the single term e-commerce is used.

E-commerce implementation follows the implementation steps in earlier chapters. That is, you really have to understand your business including your supply and selling channels. You must also know a lot about your business processes. To implement e-commerce successfully, you must involve top management as well as a substantial part of the organization. Then you must define a strategy for how you are going to go about implementing e-commerce. E-commerce is an example of several of the fundamental points in this book:

- To be successful, you must establish close ties and collaboration among IT and business units.
- The business process and the technology go hand in hand in getting results.
- Technology plays a major but not dominant role in implementation.

The objective of the implementation is to create an e-commerce solution that achieves some or all the following goals:

- Reduce costs through streamlining internal business processes.
- Shorten the time it takes to carry out business transactions.
- Improve relationships with current suppliers and customers.
- Reach out for new business.
- Strengthen your overall supply, distribution, and sales channels.

The scope of e-commerce is probably going to include most of your major business processes. E-commerce is not some add-on. It will end up being a new channel that will later replace the existing channels in the firm.

STEPS IN IMPLEMENTING E-COMMERCE

Given that e-commerce is probably going to be at the core of a business, it is not surprising that much of the implementation effort lies in the business sector. To get started, a company must select a project leader who is very familiar with the business and has some knowledge of technology. It is not necessary for the person to be a technical guru. Finding such a person in the IT organization is usually not feasible. Where else do you look? You could go to marketing. In fact, many firms select someone from marketing. However, it may be better to pick someone who is more familiar with the variety of business processes in the organization. Such a person will have worked in different departments and acted in different roles. Consider the operations area.

After a person has been selected, a small team can be formed. The team will consist of an IT systems analyst and a person from marketing. Several additional team members can be added from departments that are going to be involved. Rather than being just a simple implementation, e-commerce demands continuous effort and attention. The project is probably going to turn into a *program*. Think of a program as a long-running project.

The steps in implementation are as follows:

- *Step 1: Assess your business*. This step answers the basic questions of what others are doing in your industry segment and addresses internal issues and opportunities.
- *Step 2: Define your e-commerce strategy*. This step is extremely important because you will determine your long-term direction for e-commerce, set

goals, and define the scope of what will be required. Many companies do not address this step completely and thus implement a simple web site that no one visits.

- *Step 3: Determine your technology solution.* With the strategy in place, you can now identify the technologies (hardware, software, network, etc.) that will be required. You will also uncover the technology challenges that you will face, including interfacing to existing systems.
- *Step 4: Develop your implementation plan and budget.* Here you will not only develop a project plan, but also you define future phases of work after initial implementation. The plan is broader than implementing a software package.
- *Step 5: Implement your initial e-commerce solution.* This will be the first major achievement on the surface. However, underneath this, you will also have achieved measurable efficiency and effectiveness improvements in the business processes.
- *Step 6: Measure results and expand.* To gain support for the future, you must understand what has been achieved. Measurement includes going beyond e-commerce into the business processes. In many instances, the benefits of e-commerce may at first seem disappointing. The real benefits will lie in internal improvement.

STEP 1: ASSESS YOUR BUSINESS

In this step, you will take a good hard look at your internal business as well as what the competitors and other firms similar to yours are doing. Here are some actions you will be taking. These are numbered, but they can be performed in parallel.

Action 1: Review Your Current Internal Business Processes

This action is the same as the reviews and analysis conducted in earlier chapters. The difference here is that you are placing emphasis on those transactions that are potentially affected by e-commerce.

Action 2: Scan the Industry and Competition

You will spend time visiting web sites and printing pages. You will also search the Web for articles on the companies as well as the technology. You are trying to find out what the competitors offer and what leaders in similar industries are doing. When you gather data, you can prepare several tables.

- *Functions versus companies*. Here the functions supported on the Web site are given as rows. The companies in the industry are given as columns. The table entry can be a "yes" or "no."
- *Evaluation criteria versus companies*. The rows are performance and other evaluation criteria, and the columns are the companies. Examples of evaluation criteria might include general response time; appearance of the site; organization of the site; ease of doing a transaction; clarity of information presented; availability of telephone numbers, addresses, fax numbers; apparent volume of traffic; and apparent efforts to improve the site.

Action 3: Review Your Internal Systems and Technology

Here you are trying to determine how big a leap is needed to make the technology e-commerce ready. Without detailed design or systems analysis, you should walk through some sample e-commerce transactions with IT staff. Problems in interfacing between application systems and with e-commerce software should surface. Review your network. Can it handle the internet protocols and traffic, or does it require modernization?

Action 4: Conduct Surveys and Analysis of the Potential Visitor Audience

If you have a defined audience that you are after, then you should consider conducting a limited survey to find out if the people have a personal computer (PC) and whether they are connected and use the internet. If they do not have computer access and they are your target market, you can stop the e-commerce project right here and place it on hold for a few years. Or you can survey what types of functions they would like to have and use.

Action 5: Review Available Technology Products

This is a brief scan to be conducted after your review in Action 3. With the IT people, you can identify several candidates for e-commerce software that are most compatible with what you currently have internally. You can also identify potential consultants who could support implementation.

Present the Results to Management

After you have completed the five actions, you have a pretty good picture of what is going on both internally and externally. You are ready to discuss the results with management informally. The more formal presentation will occur the next step when you present the recommended strategy.

The informal presentation should include the following:

- *What other firms are doing in the industry.* Show sample web pages, list capabilities, and indicate benefits that firms have claimed. Also include an assessment of your firm's web site efforts.
- *What the challenges are in terms of the business.* Here you will highlight some of the business process, procedure, and policy issues that have surfaced that would impact e-commerce.
- *Technical barriers that will have to be overcome to implement e-commerce.* Problems with interfaces and the need to upgrade internal hardware and network resources are two examples of technical barriers that may exist.

STEP 2: DEFINE YOUR E-COMMERCE STRATEGY

This step is divided into two actions. The first action is to determine a general strategy that sets the direction. The second action is to take the selected general strategy and develop it in more detail.

Action 1: Define the General Strategy

First, you want to define several alternative general strategies to provide a balanced picture. The ones you develop are for your firm; thus, here are some examples developed for a firm that does business via customer e-commerce:

- *General Strategy 1. Do nothing.* With this strategy, you will not pursue any e-commerce application. This means that the problems and bottlenecks in the internal processes will not be addressed by e-commerce.
- *General Strategy 2. Develop a basic web site and storefront.* This is what many firms have done. It has often had little success. Visitors find that it is too difficult to move from the web page to placing an order so they resort to telephone traffic or some other channel.
- *General Strategy 3: Implement an e-commerce system that includes order processing and shipping.* Here you will make fundamental changes in internal business processes to expedite the flow of work. This in turn will increase efficiency and drive down costs. The strategy also supports providing more complete and satisfying service to visitors.
- *General Strategy 4. Develop a comprehensive web site for all customers.* This is the mother of all strategies. It includes the previous two strategies, but it goes beyond them to include almost all the things that someone might do. This strategy might include other nontraditional products and services.

It makes political sense to include the "do nothing" strategy. Use this to show the cost of falling behind. The second strategy really offers little more than the first strategy. In fact, one could argue that if faced to choose between the two, the first might be less embarrassing. It is clearly less expensive!

That leaves the last two strategies. Because you have no experience with e-commerce as yet, you really do not know what will prove popular and effective. Therefore, it is overkill to embark on the fourth strategy. The fourth strategy is suited for firms that are willing to take substantial risk and gamble on what people will do on their web site.

Action 2: Determine Specific Strategies

Having defined some general strategy, you are now prepared to define detailed strategies for the following dimensions:

- *Customers/suppliers.* What is your strategy for attracting, retaining, and building the relationships?
- *Products and services.* What range of products and services will be offered or included? How will they be priced?
- *Business processes.* What is the scope and range of what can be done with the business processes to support e-commerce? What will be included in terms of procedures and policies?
- *Organization.* What is allowable in terms of changing the organization to support e-commerce? This question affects, for example, marketing and sales incentives.
- *Technology.* What is the scope of the systems and technology that can be changed or modified to support e-commerce?

With General Strategy 3 selected, the strategy for customers was to offer incentives to use the site along with customized help and guidance in using products. There were no new products or services to be offered. All major business processes were viewed as fair game in terms of improvements. The organization was also within the scope, but making any major organizational structure change was off limits. Technology included interfaces with existing systems as well as some internal changes to these systems.

Action 3: Present the Strategy to Management

In presenting the plan to management, keep in mind that e-commerce may be foreign to them. Do not make the mistake of trying to train them in one or two presentations. This will be futile. Also, do not bother going into technology details. How do you make the presentation then? Center everything on the fact that e-commerce is just another modified form of doing business. It is another channel. Use business words and phrases. Give an example of a transaction to show how e-commerce differs from other channels.

What are you requesting when you make this presentation? You are not asking for money or implementation; instead, you are requesting approval to look at the technology and to build an implementation plan and budget. These will then be presented for approval. Why give this presentation on strategy if you are not asking for implementation? Because you have to have a defined direction. Otherwise, there are too many alternatives and you cannot build a plan.

STEP 3: DETERMINE YOUR TECHNOLOGY SOLUTION

Face up to the fact that you will have to acquire hardware, software, and network components. You will purchase off-the-shelf software tools to create the system and then you will have to customize it to fit your business. If this were not enough, you will have to interface with some of the existing legacy systems. This is a tall order, but it is doable if you have the will and the management support.

Here is a starting list of some things to consider:

- External service to support credit card processing
- Credit card interface software
- Internet software to interface with the catalog of products and services you offer
- Internet software to handle transactions, security, and access control and provide feedback, etc.
- Interface programs that will have to be written to link the internet software to existing software
- Interface software to link different existing systems (e.g., you may, have to link order processing to shipping)
- Software development tools for customizing the internet software as well as for building new software such as Wizards to help visitors
- Hardware for the Internet software. (You will probably require several servers. Make sure that these are scalable. One firm had a web site that was a great success; however, because they selected the wrong architecture, they ended up with 30 servers to meet demand!!)
- High-speed Internet access
- Internal upgrades to the wide and local area networks in the firm
- Business software for managing the customer or supplier relationship (This includes tracking what people do, what part of the site they visit, how often they visit, and links to direct mail and other channels.)
- Knowledge database to act as a repository for sales and tracking information, along with knowledge management data mining software tools

You can also link e-commerce to your customer service and order-taking areas. You can provide telephone references on the web site. Beyond this, you can purchase software that allows the visitor to set up a chat with your staff to answer questions and complete transactions. There is even software that allows two or more people to share the cursor at the visitor PC. This is helpful in completing forms.

This list is based on the following assumptions gathered from successful e-commerce implementations:

- The e-commerce application will not replace existing software. You cannot afford the time and effort to include replacement. You will have to live with what you have.
- The development work will likely be outsourced due to the limitations of available IT resources, the time it would take to come up to speed, and the time requirements to get the site up.

With all these components, you should develop an e-commerce architecture. The architecture is a specification of all components, what they do, and how they interrelate. It also identifies potential bottlenecks in the technology as well as where development of software will have to be done.

Where are your potential trouble spots? Here is a list from experience:

- The server and software that links to the Internet can be mis-sized resulting in performance problems. Try to oversize.
- If a variety of internet software packages are selected from different vendors, then you will be faced with additional integration work. Keep it to a few vendors.
- Lack of backup and redundancy. Given 24-hour per day and 7-day per week availability, you have to address what happens when a component goes down. You also have to address service when the legacy, host systems are running batch work and are not available for online transactions.
- Not only must the interface between the Internet software and the host systems for order processing be fast but it also must allow for the range of possible transactions and for potential failure.
- The interface between every two host systems must be carefully built because the base transactions must be supported with efficiency.

STEP 4: DEVELOP YOUR IMPLEMENTATION PLAN AND BUDGET

The Implementation Plan

This is too big to consider one plan. Instead, you will create separate plans and then combine these in a summary form for management. The plans might include the following:

- *E-commerce plan.* The goal of this project is the establishment of the hardware, software, and network for e-commerce and the installation and setup of the e-commerce software.
- *Customization of e-commerce software and new development.* This project is based on the work required to tailor the e-commerce software to your firm and to develop new software as necessary. It includes the requirements, design, development, testing, and implementation.
- *Interfaces.* This project is very important and includes all three types of interfaces: external (e.g., to credit card), internal between e-commerce and existing software, and between existing software.
- *Business processes.* This encompasses the changes to the business processes to match up with e-commerce. Also included here are the changes to other channels made necessary by e-commerce.
- *Organization related.* Included here are the incentives for staff to support e-commerce, as well as other organization changes, and make it part of the business.
- *System integration and roll-out.* The entire e-commerce architecture must be integrated and tested.
- *Marketing and promotion.* This plan addresses how the web site will be advertised and promoted.

It is in implementation planning where you can build a general team of many people working part time in addition to their regular work. This will be your approach to build teamwork and a collaborative, supportive atmosphere for e-commerce.

In Chapter 14, it is suggested that you create an issues database and spend a great deal of your time addressing issues. These actions are critical here. There will be many issues that cross projects. Here are some examples from past projects:

- How to mesh the roll-out schedule with the changes to the channels
- How to integrate e-commerce and customer service
- How to encourage creativity among people who are unfamiliar with e-commerce and who take a more traditional view of the business
- The extent of changes in the legacy systems that can be permitted given the schedule—that is, what functions will have to be deferred due to the schedule?

The Budget

Developing your budget is more complex than a standard software implementation. Here are some of the challenges you face:

- How do you estimate for software development without detailed requirements?
- What is the minimum set of capabilities that must be included in the initial roll-out?
- How do you estimate the effort for establishing the interfaces?
- What changes will be required in the business processes? Are they included in the budget? For example, you may have to change shipping practices, requiring the shipping department to hire additional staff.

Here are some guidelines. Begin with the hardware and network as these are most certain. Add in the software packages. With vendor and IT support, build estimates for interfaces and customization. Remember that the current software may also require internal changes in addition to the interfaces. Use this as your baseline budget. If you have to add process changes, then you will have to consult each department.

STEP 5: IMPLEMENT THE INITIAL E-COMMERCE SOLUTION

The first thing to do is to evaluate, select, and negotiate with the vendors. You can use some of the suggestions from previous chapters for this. Implementation then follows through a series of actions.

Action 1: Implement the Infrastructure

Keep in mind that this action includes upgrades to the existing internal hardware, system software, and network. You will also have to accommodate the environment for software development as well as testing. This may require additional servers and workstations. Do not think that these can be redeployed at the end of the initial development. You will be doing ongoing work. After you have installed the hardware and related components, you can install the internet software packages.

The biggest area where companies fall down here is in testing the infrastructure. It is particularly important to do load and stress testing to see how many visitors and transactions the architecture can support. Other tests are to consider response time and security.

Action 2: Carry Out Software Customization and Development

The recently installed internet software will have to be customized through tables and other setup procedures. Then you will embark on customized design and development for the e-commerce site. In parallel, you will develop and test the interfaces with the existing software and between the existing software. The actions in Chapter 6 apply here. System integration and testing play a prominent role here, too.

Action 3: Develop Policies and Modify Current Business Processes

You must modify the current business processes for order processing, warehousing, accounting, customer service, and other areas. Space does not permit any detailed methods for this (see Lientz, Rea, *How to Plan and Implement Business Process Improvement*, Harcourt Brace, 1999). It is probable that modifications to policies will have to be made in the following areas:

- Pricing of products and services
- Incentive programs for e-commerce
- Order fulfillment
- Customer dispute resolution
- Work priorities within departments that participate in e-commerce

Action 4: Implement the Incentive Program to Build Traffic

Do not have constrained thinking here. Incentives are not just electronic coupons to customers or rewards to suppliers. There are the following to consider:

- Open additional service and sales channels. Direct contact linked to Web customers.
- Reward salespeople for getting people onto the site.
- Ongoing sales commissions to salespeople who use the Web.

For example, one firm deployed a fleet of trucks to visit customer locations based on being a web customer. For example, you also have to consider incentives or disincentives with respect to other channels. For example, do you print fewer paper catalogs and reduce their distribution? Do you do different or less direct mailings? How do you change telephone ordering?

There is another part of the marketing — advertising. With what web sites will you advertise? What paper and other media ads will you create? People are not going to find out about this on their own. If you have a target audience, such as an industrial group or existing customers, you can do telemarketing or direct mailing. Look at what other companies that are doing similar things have done.

Action 5: Roll Out the Site

The site is rolled out through parallel efforts. You make the web site and all its infrastructure and support available. You change the internal processes in terms of policies and procedures. You will also embark on the promotion campaign.

In the short term, you will watch the web site closely for system performance and what people are doing as well as for how much volume the site generates. The initial period of use is a learning period for you. You will likely make some changes to the site to encourage certain products or services based on the initial experience.

STEP 6: MEASURE RESULTS AND EXPAND

Measurement of a business process and implementation were discussed in previous chapters. Some additional remarks are as follows:

- Pay attention to the total elapsed time for the transaction, not just to the initial part of the visitor interacting with the software on the site.
- Measure and compare the internal business processes with their performance prior to implementing e-commerce.
- Look at the web sites of your competitors and realistically rate your site with theirs.
- Determine the profile for what visitors do compared with existing customers or suppliers through the standard channels.

Remember, it is the overall measurements that you are after. When you start with e-commerce, it will have to grow to make money. It will take more time to determine how successful you were at attracting, retaining, and growing the customer base.

In terms of expansion, you should consider emphasizing different mixes of goods and services. You can also change ordering policies and other rules that appear to bottleneck e-commerce.

WHAT CAN GO WRONG?

- Marketing and sales become desperate for a web site. Management caves in, and the site only supports ordering and static information. There are no incentive programs and promotions in place. Any guess on how this will turn out?
- You roll out the site, but fail to keep it current. You also do not keep an eye on your competitors. Sales will likely drop off.
- As the site is used, a mass of sales data builds up. However, there has been no provision for anyone to analyze the data. The data reveals what items really sell better through this channel than others.

- You have retail locations or strong distributors. Your web site attempts to sell direct. Too bad that you did not work with the stores and distributors on how to coordinate the channels. As a result, they undercut the use of the site.

REDUCED SCHEDULE AND COST APPROACH

Implementing e-commerce is a formidable task. How can you cut down on the time and the cost? First, you should consider spending more time up front where effort is cheap and when the infrastructure has not yet been purchased. This will reduce costs later.

However, do not spend too much time in planning if it seriously delays implementation. A bookstore chain was late in launching their site because they wanted to get it right. They wanted to sell other items beside books. This delayed implementation by 6 months. Sales dropped. When they finally completed the implementation, their competitors caught up with them in 2 months.

EXAMPLES

Almost everyone cites Amazon.com and other similar web sites as real success stories. For every one of these, there are many others that have experienced problems. We examine some of these:

- Getting started too late. A leading brokerage firm in the United States believed that they could continue offering services the old-fashioned way. After all, they had a large market share and were quite profitable. When they finally awoke from their deep slumber, it was almost too late. They had to scramble to purchase companies to get into the game.
- Not thinking of the entire customer relationship. A company was created for consumers to buy books, CD-ROMS, and other items at discounted prices. The web site was a real success. However, the order processing was not linked electronically to order fulfillment, shipping, and accounting. The result was poor customer service and much higher labor costs than anticipated.
- Ignoring the channel. An electronics retailer did not open a web page. In fact, although they offered lower prices, the service in the stores was so bad that enraged customers opened their own web pages to attack the chain. The number of attacking web sites grew to 10. Did the electronics chain get the message? Apparently not because the complaints are still coming in. You should know that you are in trouble when people spend the time and money to establish a custom web site just to attack you.

- Concentrating on the glitter and the glitz, a casual apparel manufacturer developed a web site that included video clips and attempted to show the true colors of materials. Unfortunately, they did not get the ordering part right. Consumers could go to the site but could not order the goods!

Now moving from the negative to the positive, let's consider a company that makes metal components for metal working firms. They had traditionally sold to larger firms. Several years ago, they opened a catalog operation and found it to be successful in attracting business from smaller firms (those with less than 50 employees). They decided to investigate establishing a web site that would be the first in their industry. They took steps to find out about their customers. The vast majority had PCs and more than one half of these computer owners used the internet. This was encouraging. Because they were going after a new market segment even with the catalog, they strategized that they would have to offer value-added services that would encourage shop owners and personnel to access the site on a regular basis. Therefore, they decided to include the following features on the site:

- Bidding support for jobs
- Reference information on parts
- Installation and setup information
- A free forum for shops to sell used equipment on the site
- Online help

They developed a customized Wizard to help shop people work through a series of different business processes. This cost as much as the software and involved more than 120 employees to create it.

These were in addition to the standard features:

- Order placement
- Electronic mail
- Order tracking
- Feedback

To get people to use the site, they incentivized their sales staff to sign people up for the service. The salespeople were given bonus gifts for signing up a certain number of firms. In addition, the salespeople receive commissions on all internet sales from their signed-up customers.

LESSONS LEARNED

- Keep the web design simple. If you create a lot of complex graphics, it will turn people off.
- Focus on high performance and scalability. That is, make sure that the software loads quickly so people can get started from a standard modem in

less than 8 to 10 seconds. Ensure that you can increase your hardware, memory, and disk capacity easily.
- As you build the site, test it as if you were it a visitor. How easy is it to navigate around the site? After someone finds something they want to do, how long does it take for them to complete the transaction? There are many pretty web sites that attract a lot of people and yet generate few orders. Why? It took more than a dozen clicks to place an order.
- Develop a phased improvement approach for the site. Do not leave it alone.

SUMMARY

E-commerce is one of the fastest growing business areas in the United States and around the world. It represents the union of technology with business processes. Treating implementation of e-commerce like a software application is a big mistake and can lead to failure and loss of market share. E-commerce is likely to be one of the major ways that companies will use to streamline their internal processes to reduce the time and effort required for business transactions.

Some of the critical success factors for implementation include the following:

- Top management support and understanding of the impact of e-commerce
- A dedicated team, along with major participation from many different departments, with work performed in a collaborative manner
- A focus on establishing simple, scalable, and flexible solutions that can respond to the dynamics of e-commerce
- Incentivizing customers, suppliers, and employees to participate in and support the web site

WHAT TO DO NEXT

- Start researching some of the references in the appendix to uncover examples of web success and failure. Build your own lessons learned database.
- Conduct an external assessment of what other firms are doing in your business segment and what your suppliers or customers are doing.
- If you already have a web site, look at it critically. What functions have been developed that add value to the visitors? How has the web site been maintained? How has the site been measured in terms of impact?

Chapter 16

Overcome Implementation Issues

INTRODUCTION

Common issues and reasons for implementation failure are addressed here. These apply to all steps and actions of the book. For each issue, the following topics are covered:

- How the problem arises
- The impact if the problem is not resolved
- How to prevent the problem
- How to turn the situation around if it occurs

LACK OF TOP MANAGEMENT COMMITMENT

- **How the problem arises**
  People assume that if they present a good case for implementing a new system, they warrant top management commitment. They also sometimes believe that once the commitment is given, it remains steadfast. This may not always be true. Situations change in the business. There are changing priorities. Even with multimillion expenditures at the start of a project for a software package, the commitment should not be taken for granted.

- **The impact if the problem is not resolved**
  If top management commitment begins to fade, people pick up the signs fast. Resources are gradually or suddenly withdrawn from the project. Morale starts to sink. If the project manager goes back to management to

gain new approval and commitment, the assurances may be noncommittal. The project may be over.

- **How to prevent the problem**
Think of management commitment in terms of self-interest. What is in it for upper management if the technology is successfully implemented? Base the entire project on the new business process. The new business process is in the business department's self-interest. The process is important to the business. Therefore, management is more likely to support the project.

   After initial approval, do not take anything for granted. Assume that the commitment is tentative and will evaporate if there is a lack of progress or there are persistent, unsolved problems. Keep top management informed directly and through the business department. Never communicate in technical terms—always stick to the business aspects of the system and the process.

- **How to turn the situation around if it occurs**
If you sense that there is a problem with management commitment, then you should assess the implementation and determine where you are. If there are problems and issues unsolved, you should work on these. Next, work with the business department to better inform management of what is happening. Stay in close touch with management. Portray yourself as positive neutral. That is, you think that the project will be successful and that there will be substantial benefits. Remember that if you act as a strong advocate, management may not be as open with you.

FAILURE TO GAIN USER COMMITMENT

- **How the problem arises**
In the old life cycle, you first obtain user acceptance of requirements and then design. Many people used to interpret this as commitment. It is acceptance; it is not commitment. If you do not involve the business department, then they become detached. The problem occurs because the analyst or project leader makes assumptions about commitment. They think that it is logical that there should be commitment. But the department manager and staff have many other things on their minds. They are trying to meet business goals. They may see the implementation as a sideshow and a distraction.

- **The impact if the problem is not resolved**
User commitment is the basis for the project. If you lose this, then top management commitment may begin to waver. User commitment must

occur at all levels—not just at the top of the department. Remember that it is the people at the bottom who can make or break the implementation.

- **How to prevent the problem**
  Use the approach defined throughout—involve the business department in almost all nontechnical activities and decisions. Center all issues on the business process and the impact on it. Through involvement of many different business staff, you have a better chance to gain commitment.

- **How to turn the situation around if it occurs**
  If there is a problem with user commitment, then you should work immediately to improve communications with all parts of the department. You and the rest of the team should spend more time in the department. Involve the business staff in more activities.

MISUNDERSTANDING REQUIREMENTS

- **How the problem arises**
  Part of this stems from the myth that, if you get the requirements right, everything will be okay. Requirements are fuzzy. They change over time. Requirements are created by interpretation and often not based on fact. It can happen easily. A manager indicates that some feature or capability is a requirement. The systems analyst interprets this to the project team, including the programmers. Then the implementation has nothing to do with the requirement. Another problem arises when there is no validation or justification for requirements. You have to justify these.

- **The impact if the problem is not resolved**
  The system does not match up to the requirements. The system goes into operation. The department sees that there are problems and shortcomings. They start defining workarounds and constructing shadow systems. When later interviewed, they may indicate that the system did not deliver the goods.

- **How to prevent the problem**
  Misunderstanding requirements is a constant threat. That is why all requirements were based in the new business process as opposed to what was stated in an interview. Try to avoid paying attention only to requirements. Instead, focus on how the requirements have an impact on and are formed by the new business process.

- **How to turn the situation around if it occurs**
  If the requirements begin to get fuzzy or if there is a disagreement over requirements, consider stopping the project. Then go over all requirements

and relate these to the new business process. This is a good idea even if things are going well.

LACK OF USER INVOLVEMENT

*   **How the problem arises**
    This problem often arises from the use of technical tools that are cryptic and difficult to understand. Is it any wonder that business staff return to their departments with looks of bewilderment and confusion? Some IT believe that when they involve business staff, the project pace slows down and they have to educate the users. Users and IT alienate each other.
    There is another cause related to the business department. If the department staff are stretched thin or have tight deadlines, then they are naturally reluctant to get involved.

*   **The impact if the problem is not resolved**
    Without user involvement, the project is basically doomed. The IT group and one or two users feel increasingly isolated as time goes on.

*   **How to prevent the problem**
    User involvement has already been addressed. Another comment here is that you must make the effort to involve as many business staff as possible. This will gain you wider support.

*   **How to turn the situation around if it occurs**
    Seek out involvement of lower-level, junior staff. These people can be more readily detached from the business department's daily activities. You will have to build business involvement step by step. Also be sensitive to the schedule and pace of work in the department so that when there is peak activity, you can minimize user involvement to accommodate the department.

FAILURE TO MANAGE END-USER EXPECTATIONS

*   **How the problem arises**
    People see new systems and are besieged by the benefits of new technology. It is no wonder that user expectations are raised. Vendors and consultants can fuel the expectations even more. Then when they see the prototype or the system, they are disappointed. Another contributing factor is that the users may not be involved in the project very actively.

- **The impact if the problem is not resolved**
  High expectations are not necessarily fatal; however, there will likely be problems with acceptance of the system after implementation if expectations are not met. Things should gradually settle down as they use the system in the process.

- **How to prevent the problem**
  Downplay any expectations. Indicate that there may be no benefits if the business department does not play an active role in and work to implement the new business process. Turn the expectations toward the business process and away from the technology. The technology only enables the process, it is not the process.

- **How to turn the situation around if it occurs**
  If there are false expectations, then you should return to the new business process. Focus on expectations around the business process. What do they want to see that is different? This will turn the focus on the process.

CHANGING SCOPE, OBJECTIVES, AND REQUIREMENTS

- **How the problem arises**
  With any project that goes on for many months, there is a problem with "scope creep." People begin to assume that the system can do more than it should. Changes can occur gradually. Incrementally, small additions are made to the design and requirements.

- **The impact if the problem is not resolved**
  The scope and objectives can change to the extent that it is impossible for the system to meet these demands. There is a natural letdown. The business staff must now address the items in scope that were not addressed through workarounds.

- **How to prevent the problem**
  You begin with implementing a change control method. Warn everyone on a regular basis of the danger of expanding scope and objectives. When changes are proposed, surface the change as an issue. Then give it the proper visibility. Point out the impact on the cost, schedule, and risk of the project. Track any changes within the schedule and plan.

- **How to turn the situation around if it occurs**
  If the scope or objectives change, then you must respond quickly. If necessary, stop the work on the project. Make an overall list of requirements

that are based on the original purpose and scope. Then create a new table. The columns are: (1) added element of scope and objectives, (2) new requirement, (3) impact on cost, (4) impact on schedule, and (5) risk. Now you are ready for a management review.

LACK OF REQUIRED KNOWLEDGE AND SKILLS IN THE TEAM

- **How the problem arises**
  People were taken into the team based on what was told to you by a manager. You then find out that they do not have the skills or experience to deal with the work. Because they were assigned to the project full time, you are stuck.

- **The impact if the problem is not resolved**
  The wrong people can drag down the team. It only takes one or two "turkeys" to affect progress and morale. The rest of the team then views the project leader as a dummy because they cannot cope with the problem.

- **How to prevent the problem**
  The basic recommendation is to take people on for a series of small, immediate tasks. There should be no long-term commitment until you see what they can do. If they know that they are being tested, they will make a greater effort to perform.

- **How to turn the situation around if it occurs**
  If you have someone on the team who lacks knowledge, see if you can identify tasks that they can perform. That is why there is such an emphasis on defining business rules and transactions for the new process. This is detailed work that does not require technical knowledge. Moreover, even if the person lacks detailed, in-depth business knowledge, they can get help from others in the department.

NEW TECHNOLOGY

- **How the problem arises**
  Technology is constantly advancing. There is a temptation to employ it at the start of the project as well as during the project. The technology may promise substantial benefits, including reduced cost and schedule. People become mesmerized by these benefits, and neglect the hassle and pain of learning, using, and becoming adept with the technology.

- **The impact if the problem is not resolved**
  If new technology is inserted in the implementation, the project inevitably slows down. Time is needed for training and familiarization. Then people have to figure out how to use it. How much of what has already been done will have to be torn up? How much work has to be redone?

- **How to prevent the problem**
  At the very beginning of the new business process analysis (Chapter 3), consider available technologies and trends. If there is a significant new technology just around the corner, then perhaps the project should wait until the technology is available. Take a very conservative approach. Once the implementation has started, be very reluctant to introduce anything new. Make the new technology an issue for both management and the team to address. Weight the benefits against the cost, risk, and disruption.

- **How to turn the situation around if it occurs**
  If a new technology is introduced and you come onto the project, consider putting it on hold. There is nothing wrong with doing the other work in the project and reviewing the technology on the side. Consider developing guidelines for the use of the technology and identifying a technology expert before you start to use it.

INSUFFICIENT OR INAPPROPRIATE STAFFING

- **How the problem arises**
  Staffing problems often occur because of the approach management employs for project teams. That is, they get the team assembled at the start of the project. This is a big mistake. You really do not know how many people you might need or what skills you require. Moreover, as the implementation progresses, the requirements for staffing and the mix of staff change.

- **The impact if the problem is not resolved**
  Having inappropriate staff on the implementation can mean that critical tasks are not performed or are done below an acceptable standard. Having insufficient staff often means that there will be tasks undone, even if everyone works very hard.

- **How to prevent the problem**
  Identify a small core full-time team at the start. As the implementation proceeds, add people on a part-time basis to work on specific tasks. When they complete their work, then turn them back to their home departments. Be willing to consider adding consultants if you believe that you are going to be understaffed. Allow time to get the consultants and bring them up to speed.

- **How to turn the situation around if it occurs**
  For people who do not fit the project, begin the process to get them out of
  the project. In areas where you need help, focus only on the near-term tasks
  over the next 3 months. Then begin planning ahead for a 3-month time
  horizon. After all, you do not want to get people on the project too early.

CONFLICTS AMONG USER DEPARTMENTS

- **How the problem arises**
  There are many natural conflicts among departments due to the nature of their
  work and their mission. Examples are sales and warehousing, internal audit,
  and other groups. The problems often get worse due to personal conflicts
  among managers. These conflicts should always be anticipated.

- **The impact if the problem is not resolved**
  If department managers cannot see eye to eye, then this will have an impact
  on getting people assigned to the project. There will also be delays in
  getting issues resolved. A third area of impact occurs when you attempt to
  implement interfaces among departments.

- **How to prevent the problem**
  Anticipate the problems and discuss them openly with each department
  manager individually. Indicate that you understand that there can be
  different points of view. Also point out that the new process is in
  everyone's self-interest. As you work on issues, make sure you work
  with department managers individually.

- **How to turn the situation around if it occurs**
  If a disagreement surfaces, go to upper management only as a last resort. It
  is better to work with each manager one on one to work out a solution.
  Then implement the solution and actions in each department separately.
  Minimize the number of meetings that are attended by both departments.

MANAGEMENT DIRECTS THE IMPLEMENTATION

- **How the problem arises**
  Sometimes due to business or vendor pressure and marketing, management
  may direct demand that a specific solution be implemented. They order
  implementation without analysis of impacts or benefits. The team is
  quickly formed and typically overstaffed because it is given a high
  management priority. People start working, but there is no clear operational

goal. What is the technology intended to accomplish? What are the benefits? Rather than question management, they plunge ahead.

- **The impact if the problem is not resolved**
One outcome is that management is drawn to some other topic and loses interest. The project fails under its own weight. Unfortunately, there are many side impacts. First, managers and staff are given the signal that what they consider to be priorities is not really valid. Second, resources are taken from other areas and projects that are more important and put on this doomed project.

- **How to prevent the problem**
Keep in touch with management about what they are interested in. Stress tangible benefits. You might consider recommending that all ongoing implementations be assessed for benefits. When a technology or systems idea comes up, you are then ready for the benefit evaluation. This will slow down the rush to start implementation.

- **How to turn the situation around if it occurs**
If the snowball of implementation goes too far, you cannot do anything. You have to "nip it in the bud" by suggesting that an implementation plan be developed. This will give you time to determine benefits and requirements.

A SYSTEM IS IMPLEMENTED, BUT THERE IS NO BENEFIT

- **How the problem arises**
This is an all too frequent occurrence. The work was justified with benefits. The implementation was successfully carried out. No benefits resulted. Why? Here are three potential reasons:
  - Management failed to insist on measuring benefits after implementation.
  - Because there were no known benefits, they were shoved under the rug after implementation.
  - To get the benefits, politically unacceptable steps would have to be taken.

- **The impact if the problem is not resolved**
After a few instances of this, people fail to take benefits seriously. Like the weather, everyone talks about it but no one does anything or insists on it. Implementation priorities then shift more to politics.

- **How to prevent the problem**
There must be a firm rule by management that tangible benefits will be taken seriously. Every completed implementation should be measured. If

benefits were not achieved or were less than expected, then there should be an analysis performed to determine why this happened, what can be done about it now, and how to prevent it in the future. An alternative is to take benefits out of the picture entirely and admit that the decisions are political.

- **How to turn the situation around if it occurs**
  The analysis suggested in prevention is probably the best solution. See if the process can be changed more to obtain benefits.

MANY PROBLEMS BUILD UP DURING IMPLEMENTATION

- **How the problem arises**
  The implementation begins, but problems come up and never seem to be resolved. This can happen if there was a lack of analysis at the start. The scope grew quickly to its natural boundaries. This occurs when you try to address one legacy system, for example, and then find that there are other systems and process issues that must be faced.

- **The impact if the problem is not resolved**
  The project team and leader were not prepared for the magnitude of the problems. The project will typically become paralyzed or languish due to people waiting for decisions to be made.

- **How to prevent the problem**
  Analysis performed up front to determine the purpose and scope is one key to prevention. A second is to have a method in place to address issues.

- **How to turn the situation around if it occurs**
  The best approach is to make a list of outstanding problems and issues. Next, try to pin down the causes of the problems and their nature. Prepare a small presentation and give this to individual managers. Indicate that you have no stake in the outcome, but that decisions need to be made.

WRONG TECHNOLOGY SOLUTION CHOSEN FOR IMPLEMENTATION

- **How the problem arises**
  Even with all the analysis in the world, you can make a mistake in selecting the wrong technology. One cause is that you were faced with selecting

among the lesser of evils—that is, acceptance of undesirable choices. Perhaps, the project should have been put on hold, but it was not. You can also uncover a hidden interface issue. That is, it is mathematically possible that you will be a first-time user of some combination of technologies. Atlas Bank selected a software package that seemed to interface with all its hardware and systems software. It turned out that one of the legacy systems and the new systems were not compatible in terms of communications.

- **The impact if the problem is not resolved**
  If the implementation depends on this technology for success, then there is a big problem. Atlas Bank had to create a custom interface with their legacy system using the vendor of the software package. This cost 1 year and more than $200,000. The implementation was placed on hold.

- **How to prevent the problem**
  The only good advice is to perform extensive analysis at the start. You should try to find some customer of the technology who has a similar architecture. However, similar to Atlas Bank, any interfaces to older systems will be unique and a source of problems. Knowing this in advance, you should concentrate on any areas of risk such as these. The platform, operating system, and database management systems are of less concern.

- **How to turn the situation around if it occurs**
  Before you sign up with a vendor, find out about their resource availability to help you out if there is a problem. When the technology does fail, then you must show the vendor the seriousness of the situation. If you have to stop the project, then do it.

NEW PROCESS AND SYSTEM ARE OPERATIONAL, BUT DYSFUNCTIONAL WITH THE ORGANIZATION

- **How the problem arises**
  The system implementation was successful. The new business process does work but not well. It turns out that the organization was not changed to adapt to the new business process. People have the same job titles and are doing the same work they did before. However, they are not doing the process or using the system in the correct manner. They may be signing as two different users to do the transactions in the old way. They may still be filling out manual sheets and logs. The supervisors may not be using the system in the way intended. Parameters to route work to individuals based on different skill levels may be improperly set.

- **The impact if the problem is not resolved**
The new process will deteriorate quickly. What will be blamed? The process and system will be blamed, of course. The response may be to request that IT modify the system so that the old process can be used.

- **How to prevent the problem**
When you are defining the new process in Chapter 3, you should consider starting to define job functions and duties. You can also define new job roles based on the new process. Then when the system is defined and implementation begins, you can refine what you have and prepare people for the new process through training.

- **How to turn the situation around if it occurs**
The first thing is to measure the new process and the new system after implementation. This will surface some of the organizational issues. Make sure that you categorize these as procedural issues. If you label them as organizational or structural, then you will be in trouble and will probably have politicized the situation.

PICKING UP THE PIECES WHEN THE IMPLEMENTATION FAILS

- **How the problem arises**
Failure can occur with any implementation. Some of the reasons for failure have already been addressed. Here the situation is that failure has occurred and the question is, "What do you do about it?" Many firms will attempt to hide failure and just go on to other work. The pieces are not picked. People are reassigned and no one talks.

- **The impact if the problem is not resolved**
If nothing is done after a failure, there are no lessons learned. Furthermore, some people may think that the organization does not care about successes or failures since there is no action taken with failure. This has a dangerous impact on morale. Several times in projects this problem had to be confronted to try to improve the motivation of the staff.

- **How to prevent the problem**
There should be a defined approach to gather lessons learned from all implementation projects whether they succeed or fail. All implementation projects should have a contingency that identifies what can be done if the project work is stopped for whatever reason.

- **How to turn the situation around if it occurs**
  For failed projects, you must also perform salvage to determine what you can retain to implement something. Most frequently, you can improve the manual procedures around the current business process and system. In some instances, the hardware and network infrastructure are improved so that they can be employed to some advantage as well.

SUMMARY

This last chapter has addressed a number of issues that you may encounter in system and technology implementation. This is appropriate given that one of the themes has been to use the projects and experience of the past to help guide your present and future implementation projects.

What has been developed and presented in these chapters has been a practical approach for you to address technology implementation step by step. Each action and step have been considered in terms of how to do it with minimal time and how to avoid failure and problems. With technology ever changing, how will the method and approach here need to be updated? Experience shows that the answer is "Not that much". Specific tools and techniques improve, especially in e-commerce and knowledge management, but the underlying steps and actions remain the same. They are, in the final analysis, common sense.

The focus of the guidelines has been to reduce time in implementation. Project after project, both successes and failures, reveal that time not money or resources is the greatest enemy in successful implementation. You can obtain more money and get more and better people, but time is immutable.

Index